FM 17-74

DEPARTMENT OF THE ARMY FIELD MANUAL

M26 PERSHING MEDIUM TANK

CREW DRILL, SERVICE OF THE PIECE,

AND STOWAGE

FIELD MANUAL

> **RESTRICTED. DISSEMINATION OF RESTRICTED MATTER.**—No person is entitled solely by virtue of his grade or position to knowledge or possession of classified matter. Such matter is entrusted only to those individuals whose official duties require such knowledge or possession. (See also AR 380–5.)

by **DEPARTMENT OF THE ARMY   NOVEMBER 1949**

©2013 Periscope Film LLC
All Rights Reserved
ISBN#978-1-937684-48-8
www.PeriscopeFilm.com

DISCLAIMER:

This document is a reproduction of a text first published by the Department of the Army, Washington DC. All source material contained herein has been approved for public release and unlimited distribution by an agency of the US Government. Any US Government markings in this reproduction that indicate limited distribution or classified material have been superseded by downgrading instructions promulgated by an agency of the US government after the original publication of the document No US government agency is associated with the publication of this reproduction. This manual is sold for historic research purposes only, as an entertainment. It contains obsolete information and is not intended to be used as part of an actual training program. No book can substitute for proper training by an authorized instructor.

©2013 Periscope Film LLC
All Rights Reserved
ISBN#978-1-937684-48-8
www.PeriscopeFilm.com

DEPARTMENT OF THE ARMY FIELD MANUAL
FM 17-74

This manual supersedes FM 17-74, 11 December 1944, including C1, 3 July 1947

CREW DRILL
SERVICE OF THE PIECE,
AND STOWAGE
MEDIUM TANKS, M26 AND M45

DEPARTMENT OF THE ARMY • NOVEMBER 1949

RESTRICTED. DISSEMINATION OF RESTRICTED MATTER.—No person is entitled solely by virtue of his grade or position to knowledge or possession of classified matter. Such matter is entrusted only to those individuals whose official duties require such knowledge or possession. (See also AR 380-5.)

United States Government Printing Office
Washington : 1949

DEPARTMENT OF THE ARMY

WASHINGTON 25, D. C., *1 November 1949*

FM 17–74 is published for the information and guidance of all concerned.

[AG 300.5 (3 Aug. 49)]

BY ORDER OF THE SECRETARY OF THE ARMY:

J. LAWTON COLLINS
Chief of Staff, United States Army

OFFICIAL:
EDWARD F. WITSELL
Major General
The Adjutant General

DISTRIBUTION:

Tech Sv (2); Arm & Sv Bd (1); AFF (40); OS Maj Comd (2); MDW (11); A (ZI) (10), (Overseas) (5); CHQ (5); D (10); B (5); R 6, 7, 17, 71 (5); Bn 7 (5), 17 (25), 71 (5); C 7 (5), 17 (25), 71 (5); Sch (50); PMS&T (1); SPECIAL DISTRIBUTION.

For explanation of distribution formula, see SR 310–90–1.

CONTENTS

	Paragraphs	Page
SECTION I. General	1–2	1
II. Crew Composition and Formations	3–4	1
III. Crew Control	5–12	3
IV. Crew Drill	13–18	11
V. Service of the Piece	19–34	20
VI. Mounted Action	35–40	32
VII. Dismounted Action	41–51	43
VIII. Evacuation of Wounded from Tanks	52–56	56
IX. Inspections and Maintenance	57–62	61
X. Sight Adjustment	63	82
XI. Destruction of Equipment	64–71	83
XII. Stowage	72–74	92
APPENDIX I. Preparation of Subject Schedules		93
II. Stowage List Medium Tank, M26		102
III. Stowage List Medium Tank, M45		126
IV. References		150
INDEX		152

RESTRICTED

This manual supersedes FM 17-74, 11 December 1944, including C 1, 3 July 1947

SECTION I

GENERAL

1. PURPOSE. This manual is for the guidance of platoon leaders and tank commanders in training crew members with a view to attaining efficient teamwork in the crew operation of the Medium Tanks, M26 and M45. It is emphasized that the drills described in this manual are for the development of crew teamwork in the fighting operation of the tanks and that the ultimate goal to be attained is successful operation of the tanks on the battlefield.

2. SCOPE. This manual sets forth methods for developing speed and precision in the execution of mounted and dismounted action and service of the piece; also, procedures for inspection of the tanks and their equipment, stowage of the tanks, evacuation of the wounded, and destruction of equipment. Except where either the M26 or the M45 is indicated specifically, instructions in this manual apply to both the Medium Tank, M26 and the Medium Tank, M45.

SECTION II

CREW COMPOSITION AND FORMATIONS

3. COMPOSITION. The tank crew is composed of five members designated as follows:

Tank Commander___ TANK COMMANDER
Gunner _____ GUNNER
Bow gunner_____ BOG
Tank driver_____ DRIVER
Cannoneer_____ LOADER

4. FORMATIONS. a. Dismounted posts. The crew forms at attention in one rank (fig. 1). The tank commander takes post 2 yards in front of the right track, facing to the front. The gunner, bow gunner, driver, and loader, in that order, take posts on line to the left of the tank commander at close interval.

b. Mounted posts. (See fig. 2.) The crew forms mounted as follows:

Figure 1. Medium tank crew, dismounted posts.

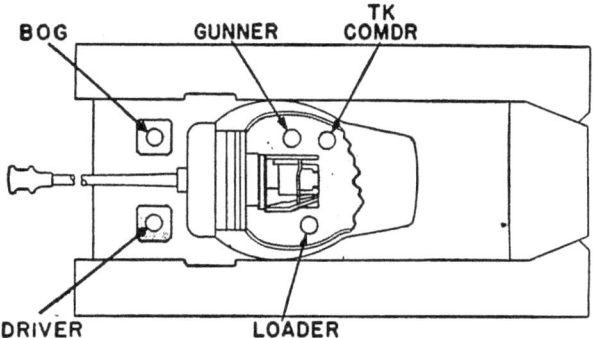

Figure 2. Medium tank crew, mounted posts.

(1) *Tank commander.* In the turret, standing on the floor, or sitting or standing on the rear turret seat.

(2) *Gunner.* In the gunner's seat, to the right of the gun (howitzer).

(3) *Bow gunner.* In the bow gunner's seat, to the right of the driver. (He is also the assistant driver.)

(4) *Driver.* In the driver's seat.

(5) *Loader.* In the turret, on the seat at the left of the gun (howitzer), or standing on the floor at the left of the gun (howitzer).

SECTION III

CREW CONTROL

5. OPERATION OF INTERPHONE AND RADIO. The tank interphone system is used for voice communication among members of the tank crew, and for communication with individuals outside the tank through the external interphone. The tank radio set is used for communication with other tanks and with other armored units. The interphone is a part of the vehicular radio set. The equipment is designed so that operation of the interphone system automatically cuts out the radio. The crew must be proficient in the operation of the interphone system if they are to obtain its maximum value in combat. Proficiency in the operation of the interphone system is gained only by continued practice.

6. PROCEDURE BY CREW MEMBERS AFTER MOUNTING. As standing operating procedure, the interphone system is checked by the crew members immediately after mounting, as follows:

  a. Tank commander.

    (1) Puts on headset, connects the breakaway cords, and inserts the microphone plug into the control box.

    (2) Prepares the radio and interphone for operation by placing switches and controls in correct operating positions, turning the set on, and selecting the desired frequency channel.

    (3) When the radio set has warmed up (about 30 seconds), he adjusts the squelch control on the receiver.

    (4) Over the interphone system he commands: REPORT.

  b. Other crew members. In the meantime, the other crew members put on, connect, and adjust their headsets and microphones. Upon hearing the tank commander give the command REPORT, each crew member answers in turn as follows:

    (1) "Gunner ready."

    (2) "Bog ready."

(3) "Driver ready."
(4) "Loader ready."

During this procedure, each crew member adjusts the volume control on his interphone control box to obtain the desired level of volume. *Care must be taken that the microphone switch does not remain in the closed position.*

Note. Immediately after the Bog reports "Ready", as described above, he dismounts, goes to the rear of the tank and checks the External Interphone, RC–298. He makes this check by opening the door to the external interphone box, removing the handset, depressing the microphone switch, and transmitting, "Commander—this is—external interphone—ready."

7. CONTROL BOX POSITIONS. Interphone control box positions are as follows:

a. Tank commander. On right wall of turret next to gunner's control box. He controls his transmission by manipulating the switch on his control box, marked RADIO-INT, to the type of transmission desired.

b. Gunner. On the right wall of turret to his right.
c. Bow gunner. On the hull cross-wall behind him.
d. Driver. On the hull cross-wall behind him.
e. Loader. On the left wall of turret to his left.

8. SETTING OF SWITCHES. a. The RADIO-INT switch on the interphone control box of each crew member, except the tank commander, is ineffective when modified in accordance with MWO SIG 11–600–9, and may be in either the RADIO or INT position. On unmodified tanks, the RADIO-INT switches of all crew members except the tank commander must remain in the RADIO position. The tank commander's switch, however, being spring-loaded, remains in the INT position, which permits any crew member to use the interphone system and all members to receive incoming radio signals. When the tank commander desires to transmit over his radio, he must first push the switch to the RADIO position (*and hold it there while transmitting*), and then depress his microphone switch and transmit. In an emergency, any crew member may interrupt the radio transmission of the tank commander by merely depressing his microphone switch; this action automatically switches the system from RADIO to INTERPHONE.

b. If the tank is equipped also with a Radio Set, AN/VRC-3, there are two changes in the above procedure.

> (1) First, the tank commander has two control boxes—one for the AN/VRC-3 and one for the SCR-508 (SCR-528). Unless the desires to receive or transmit over the AN/VRC-3, the RADIO-INT switch of the control box for the AN/VRC-3 is left in the INT position; the operation procedures of the interphone system and the SCR-508 are unchanged. If he desires to transmit or receive over the AN/VRC-3, he places the RADIO-INT switch of the control box for the AN/VRC-3 in the RADIO position. Upon doing this, he is disconnected from the interphone system and the SCR-508; he can transmit or receive only over the AN/VRC-3. It is the duty of the tank commander to monitor his radio receiver at all times except when speaking over the

interphone or transmitting over the radio. He can monitor both the radio receiver and the interphone with his switch set at the INT position.

(2) Second, the loader has a control box for the AN/VRC-3, which is the only control box that he uses. With the RADIO-INT switch of the control box for the AN/VRC-3 in the INT position, the loader cannot transmit or receive over the AN/VRC-3; he has the same interphone and SCR-508 reception facilities as the other crew members. When he places the switch in the RADIO position, he may transmit or receive over the AN/VRC-3, but he is disconnected from the interphone system. The loader is designated as monitor-operator of the AN/VRC-3.

9. RADIO CHECK. As a part of the daily before-operation inspection, the tank commander will make the following radio check:

a. Cords.
 (1) See that insulation and plugs are dry, unbroken, clean, and making good contact.
 (2) Arrange loose cordage to prevent its entangling personnel or equipment.

b. Antenna. See that—
 (1) Mast is complete and sections are fastened securely.
 (2) Leads at transmitter, receiver, and mast base are tight.
 (3) Mast base is tight and not cracked.
 (4) Insulators passing through armor plate and bulkheads are whole and in place.

c. Set mountings, snaps, snubbers, etc. Check for security and condition.

d. Microphones, switches, and headsets. Check for condition and proper position. Replace from spares, if necessary, and turn in defective items for repair or replacement.

e. Spare antenna sections. See that they are placed correctly in the roll and stowed to protect them from damage and keep them out of the way of personnel.

f. Ground lead. Check connection at both ends.

g. Fuzes. Check condition of those in use, and see that spares are of proper rating and in sufficient quantity.

h. Cleanliness. See that radio and accessory equipment is clean.

i. Crystals. Check for number, position, and frequency. Be sure required crystals are present.

10. CHECKING INTERPHONE EQUIPMENT. It is the duty of each man to check his interphone equipment upon mounting the tank; he should see that it is maintained properly, and should report any difficulties to the tank commander.

11. USE OF DEFINITE TERMINOLOGY. Terminology used by tank commanders in controlling their crews is set forth in paragraph 12. Failure to use standard, specific interphone language causes misunderstanding and disorder and is harmful to discipline. It is emphasized that adherence by all crew members to this standard language is essential to efficient operation of the tank.

12. INTERPHONE LANGUAGE. a. Terms.

Tank commander	TANK COMMANDER
Driver	DRIVER
Gunner	GUNNER
Loader	LOADER
Bow gunner	BOG
Any tank	TANK
Armored car	ARMORED CAR
Any unarmored vehicle	TRUCK
Any antitank gun	ANTITANK
Infantry	TROOPS
Machine gun	MACHINE GUN
Airplane	PLANE

b. Commands for movement of tank.

To move forward	DRIVER MOVE OUT
To halt	DRIVER STOP
To reverse	DRIVER REVERSE
To increase speed	DRIVER SPEED UP
To decrease speed	DRIVER SLOW DOWN
To turn right (left)	DRIVER RIGHT (LEFT)—STEADY...ON
To turn right (left) 180°	DRIVER RIGHT (LEFT) ABOUT—STEADY...ON

To move toward a terrain feature or reference point, tank being headed in proper direction____	DRIVER MARCH ON WHITE HOUSE (HILL, DEAD TREE, ETC.)
To follow in column_____	DRIVER FOLLOW THAT TANK (DRIVER FOLLOW TANK B-9)
To follow road or trail to the right (left)_____	DRIVER RIGHT (LEFT) ON ROAD (TRAIL)
To start engine_____	DRIVER TURN IT OVER
To stop engine_____.	DRIVER CUT ENGINE
To proceed in a specific gear _____	DRIVER THIRD (FIRST) (FOURTH) GEAR
To proceed at same speed__	DRIVER STEADY

c. Commands for control of turret.

To traverse turret___ GUNNER TRAVERSE RIGHT (LEFT)

To stop turret traverse_____ STEADY . . ON

d. Fire commands. See FM 23–100.

SECTION IV

CREW DRILL

13. DISMOUNTED DRILL. a. To form medium tank crew. Being dismounted, the crew takes dismounted posts (fig. 1) at the command FALL IN.

b. To break ranks. Being at dismounted posts, the crew breaks ranks at the command FALL OUT. Crew members habitually fall out to the right of the tank.

c. To call off. Crew being at dismounted posts, at the command CALL OFF, the members of the crew call off in turn as follows:

    (1) Tank commander____ "Tank commander"
    (2) Gunner_____ "Gunner"
    (3) Bow gunner_____ "Bog"
    (4) Driver_____ "Driver"
    (5) Cannoneer_____ "Loader"

d. To change designations and duties.

    (1) Crew being at dismounted posts, at the command FALL OUT TANK COMMANDER (GUNNER) (DRIVER)—

        (*a*) The man designated to fall out moves along the rear of the rank to the left flank position and becomes loader.

        (*b*) The crew members on the left of the vacated post move smartly to the right one position and prepare to call off their new assignments.

(c) The acting tank commander starts calling off as soon as the crew is re-formed in line.

(2) The movement may be executed by having any member of the crew fall out except the loader.

(3) All movements should be executed at double time with snap and precision.

Figure 3. Medium tank crew mounting through hatches.

14. TO MOUNT MEDIUM TANK CREW (fig. 3). Crew being at dismounted posts.

Tank commander	Gunner	Bow gunner	Driver	Loader
Command: PREPARE TO MOUNT.				
About face.	About face.	About face.	About face.	About face.
Command: MOUNT.				
Stand fast.	Mount right fender.	Stand fast.	Stand fast.	Mount left fender.
Mount right fender.	Mount right stowage box.	Mount right fender.	Mount left fender.	Mount left stowage box.
Mount right stowage box.	Enter turret hatch.	Enter hatch; take mounted post.	Enter hatch; take mounted post.	Enter turret hatch.
Enter turret hatch.*	Take mounted post. Connect breakaway plugs.	Connect breakaway plugs.	Close battery master switch.	Take mounted post. Connect breakaway plugs.
Take mounted post.			Connect breakaway plugs.	
Turn on radio.				
Connect breakaway plugs.				
Command: REPORT.				

*When gun is to the rear, tank commander crosses turret and enters by vision cupola hatch.

Tank commander	Gunner	Bow gunner	Driver	Loader
	Report "Gunner ready."	Report "Bog ready."	Report "Driver ready."	Report "Loader ready."

15. TO CLOSE AND OPEN HATCHES.

a. To close hatches. Crew being at mounted posts.

Tank commander	Gunner	Bow gunner	Driver	Loader
Command: CLOSE HATCHES.	Release turret traversing lock and insure that the turret or the cannon does not block hatches.			
Close hatch. Adjust periscope. Command: REPORT.	Close hatch. Raise periscope.	Close hatch. Raise periscope.	Close hatch. Raise periscope.	Close hatch. Raise periscope.

Tank commander	Gunner	Bow gunner	Driver	Loader
	Report "Gunner ready."	Report "Bog ready."	Report "Driver ready."	Report "Loader ready."

b. To open hatches.

Tank commander	Gunner	Bow gunner	Driver	Loader
Command: OPEN HATCHES.	Release turret traversing lock and insure that the turret or the cannon does not block hatches.	Lower periscope.	Lower periscope.	Lower periscope.
Open hatch.		Open hatch.*	Open hatch.*	Open hatch.
Command: REPORT.	Report "Gunner ready."	Report "Bog ready."	Report "Driver ready."	Report "loader ready."

Crew being at mounted posts.

* When hatch is open turn periscope and fold flat into hatch cover, to clear path of turret bulge.

16. TO DISMOUNT MEDIUM TANK CREW.
Crew being at mounted posts, hatches open, turret straight forward or straight to the rear.

Tank commander	Gunner	Bow gunner	Driver	Loader
Command: PREPARE TO DISMOUNT.				
Disconnect breakaway plugs.	Disconnect breakaway plugs.	Disconnect breakaway plugs.	Disconnect breakaway plugs.	Disconnect breakaway plugs.
Turn off radio.			Open battery master switch.	
Command: DISMOUNT.				
Emerge from turret.	Stand fast.	Emerge from hatch.	Emerge from hatch.	Emerge from turret.
Move to right stowage box.	Emerge from turret.	Move to right fender.	Move to left fender.	Move to left stowage box.
Move to right fender.	Move to right stowage box.	Take dismounted post.	Take dismounted post.	Move to left fender.
Take dismounted post.	Move to right fender.			Take dismounted post.
	Take dismounted post.			

17. TO DISMOUNT THROUGH ESCAPE HATCHES.

Without weapons, crew being at mounted posts.

Tank commander	Gunner	Bow gunner	Driver	Loader
Command: THROUGH ESCAPE HATCHES, PREPARE TO DISMOUNT.				
Disconnect breakaway plugs. Turn off radio.	Disconnect breakaway plugs. Traverse turret to give access to forward compartment.	Disconnect breakaway plugs. Drop right escape door.	Disconnect breakaway plugs. Drop left escape door. Open battery master switch.	Disconnect breakaway plugs.
Command: DISMOUNT.				
Stand fast.[1]				
Move into forward compartment.	Move into forward compartment.			Move into forward compartment.
Dismount through escape hatch.	Dismount through escape hatch.	Dismount through escape hatch.	Dismount through escape hatch.	Dismount through escape hatch.
Crawl out and take dismounted post.	Crawl out and take dismounted post.	Crawl out and take dismounted post.	Crawl out and take dismounted post.	Crawl out and take dismounted post.

---

[1] Drill with tank gun forward; when gun (howitzer) is to the rear, tank commander dismounts ahead of gunner.

① FALL IN.

② MOUNT.

Figure 4. Pep drill.

③ DISMOUNT: ON THE LEFT OF YOUR TANKS, FALL IN.

④ BY THE RIGHT FLANK. MARCH; MOUNT.

Figure 4—Continued.

18. PEP DRILL. To vary the drill routine and to sustain the interest of the crew members, unexpected periods of pep drill are introduced into the training. Pep drill (fig. 4) is a series of precision movements executed at high speed and terminating at the position of attention, either mounted or dismounted. For example, the crews being dismounted, the platoon commander commands: IN FRONT OF YOUR TANKS, FALL IN; MOUNT; DISMOUNT; FALL OUT TANK COMMANDER; ON THE LEFT OF YOUR TANKS, FALL IN; FORWARD, MARCH; BY THE RIGHT FLANK, MARCH; TO THE REAR, MARCH; MOUNT. Preparatory commands for mounting and dismounting are eliminated in this drill. Posts of all crew members are changed frequently.

SECTION V

SERVICE OF THE PIECE

19. GENERAL. a. The crew of the tank gun (howitzer) consists of the gunner, who aims and fires the piece; the loader, who loads the piece; and the tank commander, who controls and adjusts fire.

b. Training in service of the piece must stress rapidity and precision of movement, and teamwork. Smooth coordination and speed in the service of the piece may be the difference between life and death on the battlefield.

20. POSITIONS OF GUN (HOWITZER) CREW. Positions of the tank commander, gunner, and loader are prescribed in paragraph 4b and illustrated in figure 2.

21. TO OPEN THE BREECH. a. 90-mm gun. Grasp the grip portion of the operating handle, release the latch on the grip, and pull the handle to the rear and down. *Immediately return the operating handle to its upward position and latch it.*

b. 105-mm howitzer. Grasp the breech operating handle and squeeze the latch until it is disengaged from its catch. Push the breech operating handle to the rear and right as far as it will go.

22. TO CLOSE THE BREECH. a. 90-mm gun. The insertion of a round in the gun trips the extractors and causes the breechblock to close automatically. This feature makes it necessary to use care in closing the breech when the gun is unloaded. To close the breech, insert an empty shell case in the breech, base foremost, and trip the extractors. The breechblock will close, pushing the shell case forcibly upward. If an empty shell case is not available, a block of wood of the proper size may be used. (A live round should not be used because of the danger of deforming the rim and causing a stoppage later on. Likewise do not use any piece of steel, such as a wrench handle, because if it happened to slip and is struck by the closing breechblock, the extractors or chamber might be burred.)

b. 105-mm howitzer. The 105-mm howitzer is not equipped with an automatic breechblock and therefore must be closed manually. To close the breech, grasp the operating handle with the left hand and move it to the left and forward. *Check to see that the latch locks the handle in the closed position.*

23. SAFETY PRECAUTIONS. a. Before firing and during lulls in firing, the gun (howitzer) will be inspected to see that there is no obstruction in the bore.

b. The gunner should release the firing trigger of manual firing lever after firing to prevent injuring the loader.

c. The gunner waits for the loader's signal that the gun (howitzer) is loaded and that he is clear of the recoil before operating the firing trigger.

d. During range and combat practice firing, an officer will be responsible for the observation of safety precautions. After firing, the tank will remain in position and personnel will stay to its rear until an officer has inspected the gun, determined that it is unloaded, and authorized movement of the tank.

e. In loading the gun (howitzer), care must be taken not to strike the fuze or primer of a shell against any solid object; after loading, the loader must take care to remain clear of the path recoil.

f. Stuck rounds will be removed from the bore only with a bell rammer, which is made for this purpose. The method of removing is given in paragraphs 28 and 29.

g. The safety precautions prescribed in paragraph 27 and in SR 385-310-1 will be observed strictly when removing a misfire from the gun (howitzer).

h. Before ammunition is stowed and loaded, it will be inspected to see that it is clean and not deformed.

i. Fuzes will not be disassembled.

j. When the breech of the 90-mm gun is opened manually, the operating handle is returned immediately to its upward position and latched.

k. The fingers should not be used to trip the extractors of the 90-mm gun.

l. In loading, care will be taken not to dent or burr the projectile by striking it against the breech ring.

m. Tank weapons, except the antiaircraft gun, are fired only when the driver's and bow gunner's hatches are closed.

n. Any individual who observes a condition making firing unsafe will call immediately or signal CEASE FIRING.

o. Crews will cease firing immediately at the command or signal CEASE FIRING.

p. See safety regulations in SR 385–310–1.

24. TO LOAD THE GUN (HOWITZER). a. 90-mm gun.

(1) Open the breech and return the operating handle to its upward position. (Check engagement of latch.)

(2) Secure a round of ammunition; grasp it by the base of the shell case with the right hand and in the rear of the ogive with the left hand.

(3) Place the nose of the projectile in the breech recess, taking care not to strike the fuze against anything. Move the round forward until the nose of the projectile rests in the chamber; remove the left hand and push the round until it is well started into the chamber. Then, with the right fist or the heel of the right hand, vigorously push the round forward into the chamber, slightly rotating the body to the left from the hips and sliding the hand off the round above and to the left to insure clearing the breech. The au-

tomatic breechblock will push the hand clear if it should follow the round too far into the breech recess. Move to the left side of the turret, clear of the path of recoil, and signal "Ready" by tapping the gunner's left leg.

b. 105-mm howitzer.

(1) Open the breech.

(2) Secure a round of ammunition; holding it with the right hand at the base of the cartridge case and the left hand at the middle of the assembled round, carefully insert the nose of the projecticle into the chamber. Avoid striking the fuze. Remove the left hand from the round and with it grasp the operating handle. Then, with the right fist or the heel of the right hand, thrust the round home into the chamber. As the rim of the cartridge case engages the extractor, it starts the closing motion of the breechblock. When this motion is felt, close the breech by moving the operating handle to the left and forward with the left hand. *Check to see that the latch locks the handle in the closed position.* Move the body and both arms to the left clear of the path of recoil, and signal "Ready" by tapping the gunner's left leg.

25. TO LAY THE GUN. Under the guidance of the tank commander or by use of the periscope, move the field of the telescope onto the target by the quickest practicable method. To lay for direction, traverse until the center line of the telescope is on the target or until the proper lead is taken. Then move the gun until the target shows at the proper range, as indicated by its relation to the range lines of the reticle. In making final adjustment for both range and direction, the last movement of the gun must be made against the greatest resistance in order to minimize play in traversing and elevating mechanism. Thus, to lay on a target below his position, the gunner of a tank canted on a hill side will allow the gun's weight to assist him in depressing and traversing it easily downward *past* the target. He then will make final adjustment *upward against* the weight of the gun, thereby laying it firmly on target with mechanical play eliminated.

26. TO FIRE THE GUN (HOWITZER). With the right forefinger depress the firing trigger, located on the power traversing handle.* If the gun (howitzer) fails to fire, proceed in accordance with paragraph 33. The gun (howitzer) may also be fired by operating the firing lever on the right of the gun (howitzer), opposite the gunner's left shoulder, or by operating the foot switch beneath his seat.

27. TO UNLOAD AN UNFIRED ROUND OR A MISFIRE. a. To unload an unfired round the loader cups his hands close behind the breech to catch the base of the round as it emerges and to prevent it from dropping to the floor. The gunner, by means of the operating handle, opens the breech *slowly*. (*The breech must not be opened rapidly, or the case will become separated from the projectile.*) The loader then removes the round and returns it to its rack.

---
*The button under the thumb at the top of the power traversing handle is the electrical firing switch for the coaxial machine gun.

b. To unload a misfire, the following precautions will be taken. Two attempts will be made to fire the piece before the breech is opened and the round removed. Wait 1 minute from the time of the last attempt before opening the breech. Prior to removing the round, personnel unnecessary to the operation should be cleared from the vicinity. Then remove the round. In the event that the gun (howitzer) has been subjected to continuous fire for a considerable length of time before misfire, the tube will be hot, and the heat of the gun (howitzer) can cause the fuze, projectile filler, or propellant to explode. An explosion under such circumstances is called a cook-off. If the tube is hot, play water on it until it is cool, then remove the round. If water is not available, all personnel must stand clear of the gun until it is cool. Then remove the round. Rounds which misfire will not be returned to the racks, but will be removed to a safe place and turned over to ordnance personnel.

28. TO REMOVE A STUCK PROJECTILE. If the case and projectile become separated despite care in opening the breech, the chamber should be filled with rags to form a cushion so that the projectile will not damage the breechblock. The breech should be closed and the procedure described in paragraph 29 should be followed. After the projectile is free in the chamber, the breech should be opened, the projectile removed and disposed of in accordance with existing regulations. The chamber should be cleaned.

29. TO UNLOAD A STUCK ROUND. When a round is stuck in the gun and it is impossible or inadvisable to fire it out, it will be removed *under the direct supervision of an officer*, except in combat. The breech being open, the loader takes position to receive the round as it is pushed from the chamber. The bow gunner dismounts, inserts the bell rammer in the muzzle of the gun and pushes it gently down the bore until it is seated on the ogive of the projectile. Exerting a steady pressure, he shoves the round clear so that it may be removed by the loader. If the weight of several men applied to the staff does not suffice and the urgency of the situation necessitates removal of the projectile by tank crewmen, leverage may be applied by means of a 2- by 4-inch board or other suitable object connected to the tank by a rope at one end. All parts of the body should be kept as clear as possible from the muzzle or breech during the operation. If this procedure fails to remove the round, experienced ordnance personnel should be called. In combat, to avoid exposing personnel to enemy fire, the round sometimes can be pried out by using the base of an empty case as a lever.

30. PROCEDURE IN CASE OF DEFORMATION OF 90-MM APC. Use of the rammer, as prescribed in paragraph 29, to unload stuck rounds of armor-piercing capped ammunition may deform or loosen the aluminum windshield. Firing a projectile in this condition may cause serious damage to the gun or at least have a marked effect on accuracy at ranges over 1,000 yards. Hence, if it is necessary to use the round, the loose or deformed windshield should first be removed. The projectile will then have ballistic characteristics approximately the same as those of ordinary armor-piercing ammunition.

31. MALFUNCTIONS. Malfunctions of the gun (howitzer) may be divided into three general classes—failure to load, failure to fire, failure to extract. Paragraphs 32, 33, and 34 give the causes of the principal types of failure and the immediate action to be taken in each case.

32. MALFUNCTIONS—FAILURE TO LOAD.

Failure	Cause	Immediate action
Round does not fully enter chamber.	Stuck round	One or more of following: Remove obstruction from chamber. Clean dirty round. Remove bulged (deformed) round. For procedures in removing separated or stuck rounds, see paragraphs 28 and 29. Withdraw round and try again.
Breech does not close.	Round pushed home with insufficient force to trip extractors (90-mm) or clear breechblock (105-mm).	
	Bent or undersized case rim	Turn round so that rim engages extractor(s), or use new round.
	Obstruction, dirt, or friction in breech mechanism.	Remove obstruction; or disassemble, clean, lubricate, and reassemble breechblock.
	Worn or broken extractors (90-mm)	Replace.
	Weak or loose closing spring (90-mm)	Increase tension on spring to secure proper action.

858606°—49——3

33. MALFUNCTIONS—FAILURE TO FIRE.

Failure	Cause	Immediate action
Gun (howitzer) does not return to battery.	Obstruction between breech ring and rear portion of mount.	Remove obstruction. (When necessary use tank jack between breech ring and shoulder guard bracket of mount.)
If gun (howitzer) is in battery:		
Action of trigger mechanism restricted.	Safety on SAFE.	Move safety to FIRE.
Blow of firing pin fails to fire round.	Defective round.	Recock gun (howitzer) and attempt to fire a second time. If unsuccessful, remove round to determine cause of misfire. (See SR 385-310-1; also pars. 27–30 for procedure in removing live rounds.)
	Weak blow on primer due to obstruction, dirt, or friction in firing mechanism.	Remove round (see above references for procedure) disassemble firing mechanism and remove obstruction; or clean, lubricate, and reassemble.
	Broken tip on firing pin.	Replace firing pin.
	Broken or weak firing spring.	Replace firing pin.
Firing pin fails to strike primer.	Obstruction, dirt, or friction in firing mechanism.	Disassemble firing mechanism and remove obstruction; or clean, lubricate, and reassemble.

Weak or broken firing spring.	Replace.
Defective firing pin.	Replace.
Defective cocking lever.	Replace.
Defective cocking fork.	Replace.
Defective cocking lugs on percussion mechanism.	Replace mechanism.
Defective sear.	Replace.

34. MALFUNCTIONS—FAILURE TO EXTRACT.

Failure	Cause	Immediate action
Breech opens, but case is not extracted.	Broken extractor(s).	Pry or ram out empty case and replace extractor(s).
	Undersized or bent rim.	Pry or ram out.
	Obstruction, dirt, or friction in breech mechanism.	Remove obstruction; or disassemble, clean, lubricate, and reassemble breechblock.
Breech fails to open, or opens only partially.	Broken operating crank (90-mm).	Remove case. Replace crank.
	Broken operating crank cam (90-mm).	Remove case. Replace cam.

SECTION VI

MOUNTED ACTION

35. TO PREPARE TO FIRE. Crew being at mounted posts with hatches open and tank gun (howitzer) in forward position; antiaircraft gun is uncovered and half loaded, if required by the tactical situation.

Tank commander	Gunner	Bow gunner	Driver	Loader
Command: PREPARE TO FIRE.	Unlock gun (howitzer) traveling lock; release turret traversing lock.	Lower seat. Release bow gun traveling lock. Half load bow gun. Check ammunition. Clean periscope.	Lower seat. Clean periscope.	Open breech gun (howitzer). Inspect bore and chamber of tank gun (howitzer).
Clean gunner's, loader's, and tank commander's periscopes; gun telescope; and cupola vision blocks.	Check firing controls (including solenoids).			Start auxiliary engine.

Check elevating mechanism.

Check vane sight.

Check periscope and sights.*

Close hatch, if desired.

Uncover and check elevation quadrant and azimuth indicator.

Turn on, and engage power traverse; check power traverse operation; center gun (howitzer) to front.

Close hatch; raise periscope.

Close hatch; raise periscope.

Half load coaxial machine gun.

Open floor ammunition compartments, check ammunition.

Check ready rounds and machine gun ammunition.

Close hatch.

---

*The periscope already will be raised, having been adjusted previously for the day. Once adjusted, the periscope remains raised; lowering it may disturb the adjustment.

Tank commander	Gunner	Bow gunner	Driver	Loader
	Start and check operation of gyro-stabilizer (return howitzer to manual elevation unless ordered otherwise). (On 105-mm howitzer only.)			
Command: REPORT.	Report "Gunner ready."			
		Report "Bog ready."	Report "Driver ready."	Report "Loader ready."

36. DUTIES IN FIRING.

Tank commander	Gunner	Bow gunner	Driver	Loader
Give fire commands (FM 23–100).	Lay gun (howitzer) for deflection and range.	Fire on designated targets and on emergency targets as they appear.	Turn on forced ventilation during periods of firing.	Load the type of ammunition indicated in fire command. (Inspect each round; change fuze setting if ordered.) Close breech (howitzer).
Observe and sense each round and correct gunner by subsequent fire commands.	Fire on targets designated. Continue to fire as directed.	When not firing, observe in assigned sector.	Operate tank at constant speed and as smoothly as possible.	Signal "Ready" each time gun (howitzer) is loaded by tapping gunner on left leg. Reload all turret weapons.

35

Tank commander	Gunner	Bow gunner	Driver	Loader
Control driver over interphone.	Call "Misfire" if cannon fails to fire.		Steer only to change course as directed by tank commander, or to avoid obstacles. When about to pass over rough ground or change course, warn gunner by calling "Rough" or "Changing course."	See that all fuzes are SQ (superquick) unless ordered otherwise. In case of misfire, check that breech is closed, gun (howitzer) in battery; recock gun (howitzer) and signal READY to Gunner.
	Call "Stoppage" if coaxial gun fails to fire.		During lulls in normal activity observe in assigned sector.	Reduce stoppages in coaxial machine gun.

Fire coaxial gun by hand when directed by gunner. Fire AA gun.

Watch recoil and counterrecoil of gun for malfunctions.

Keep record of ammunition expended for entry in gun book by platoon leader (number of rounds of each type).

Tell loader when to fire coaxial gun if solenoid fails to operate.

When unable to see designated target, notify tank commander and lay gun (howitzer) for deflection and elevation, using auxiliary fire control instruments. Fire gun (howitzer) on command. Make designated corrections in deflection and elevation.

Assist loader fire AA gun.

When gunner is unable to see target, adjust fire by modified direct fire methods.

37

Tank commander	Gunner	Bow gunner	Driver	Loader
	During lulls in firing, observe in assigned sector.			Inform tank commander when necessary to restow ammunition. During lulls in normal activity observe in assigned sector.

37. TO SECURE GUNS.

The drill described below includes minimum operations required to put the tank in proper condition for movement after it has been prepared for combat or after range practice. If time permits, additional operations are performed, that is, the gunner checks sight adjustment and covers the elevation quadrant bubble and the azimuth indicator. The tank commander also may order that the bores of all weapons be swabbed.

Tank commander	Gunner	Bow gunner	Driver	Loader
Command: (CEASE FIRING) SECURE GUNS.	Turn off firing switch; center and elevate tank gun (howitzer).	Clear bow machine gun; engage traveling lock.		Clear coaxial machine gun. Clear tank gun (howitzer); inspect bore and close breech.

Command					
Open hatch------	Turn off power traverse, engage hand traverse. Check sight adjustment. Lock gun (howitzer) in travel position. Turn off gyrostabilizer (on M45, if operating). Engage manual elevating mechanism. Lock turret lock.	Check position of turret weapon; open hatch. Raise seat to convoy position.	Check position of turret weapons; open hatch.* Raise seat to convoy position.		Shut off auxiliary generator. Open hatch and clear AA machine gun.
Command: REPORT.	Report "Gunner ready."		Report "Bog ready."	Report "Driver ready."	Report "Loader ready."

*When hatch is open, turn periscope and fold flat into hatch cover to clear path of turret bulge.

38. TO LOAD ALL WEAPONS. The tank gun (howitzer) is loaded on command. Normally, this is the fire command, but some types of action may make it necessary to load prior to the appearance of a target. Machine guns are clear until the command is given **PREPARE TO FIRE**, whereupon they are half-loaded. When the fire command is given or the unit is deployed for combat, all machine guns will be loaded fully. This does not apply necessarily to the antiaircraft gun, which is half loaded in accordance with tactical requirements.

39. USE OF AMMUNITION. *a.* The order of withdrawing ammunition from its stowage space in the tank (fig. 5) is based on the principle of saving the rounds in the ready racks for emergency use. With the gun of the M26 tank to the front, the loader has access to the three floor compartments one, two, and four on the left (fig. 5 ①) and compartment three just right of the center of the turret (fig. 5 ①), as well as the rounds in the ready rack. The gunner's seat covers the other two floor compartments five and six on the right (fig. 5 ①), and the turret must be traversed 90° right or left to reach the racks against the left and right wall of the turret, respectively. With the howitzer of the M45 tank to the front, the loader has access to the four floor compartments, one, two, three, and four on the left (fig. 5 ②), and the rear compartment five right of the center of the turret (fig. 5 ②), as well as the rounds in the ready rack. The gunner's seat covers the other three floor compartments six, seven, and eight on the right (fig. 5 ②) and the turret must be traversed right or left to reach these racks. In combat the turret will not always point directly forward, and the accessibility of ammunition may be altered considerably as a result.

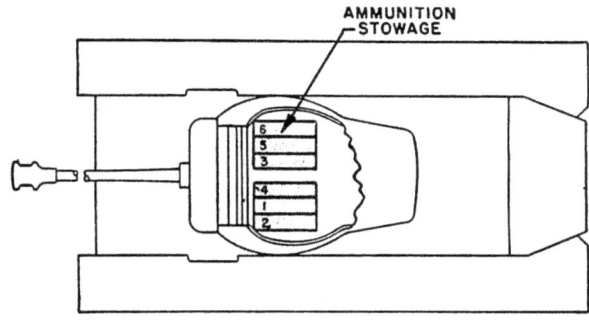

① 90-mm ammunition.

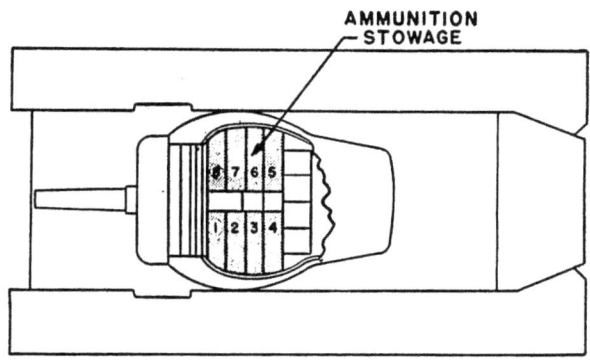

② 105-mm. ammunition.

Figure 5. Ammunition stowage.

b. Each ammunition compartment is emptied completely, if practicable, before the next is opened for use. On the M26 tank, the rounds in the ready rack and those in the right-hand compartment four, left side of turret, are kept as a reserve for action where speed in loading is of the utmost importance. On the M45 tank, the rounds in the ready rack and those in the two rear compartments three and four, left side of turret, are kept as the reserve. As time permits, or on the command RESTOW AMMUNITION, the loader moves rounds from the least accessible racks and then from the compartments on the right side of the turret to those which have been emptied in firing.

c. Upon completion of restowing, a report is made of the number of rounds of cannon ammunition remaining in the tank. For example, the loader reports: "Two two rounds AP, three one HE remaining."

40. TO LOAD AMMUNITION. Ammunition for the tank gun (howitzer) should be loaded and stowed with great care to prevent striking the fuze end or the primer on a hard surface, burring the rotating band, or denting the case (see TM 9–1900). Each round is inspected for imperfections and cleanliness prior to stowing it in the tank. Fuze settings of HE rounds are inspected also to see that they are set at SQ (superquick). Both cannon and machine-gun ammunition will be passed through the hatches or the pistol port; a man stationed on the hull or fenders relays it to crewmen in the tank. Since the fuze of the projectile is bore safe, it is less sensitive than the primer. Therefore, in order to prevent detonation of the primer in case a round should be dropped, ammunition should be passed with fuze down.

SECTION VII

DISMOUNTED ACTION

41. TO FIGHT ON FOOT, DISMOUNTING THROUGH HATCHES. (See figs. 6 and 7.) **a.** (Crew at mounted posts, with tank gun (howitzer) forward, and hatches open.) Crew members, including the tank commander, remain below hatches while preparing to dismount and fight on foot. Having completed their preparations, they maintain their positions until the command DISMOUNT is given.

Tank commander	Gunner	Bow gunner	Driver	Loader
Command: PREPARE TO FIGHT ON FOOT.				
Disconnect breakaway plugs.	Disconnect breakaway plugs.	Disconnect breakaway plugs. Procure grenades.	Disconnect breakaway plugs.	Disconnect breakaway plugs. Procure grenades.
Order distribution of grenades.	Pass three boxes caliber .30 ammunition to loader.	Remove traveling lock from bow gun.		Receive ammunition from gunner.
Take grenades, binoculars, and stand fast.				

43

Tank commander	Gunner	Bow gunner	Driver	Loader
		Dismount bow gun and attach elevating mechanism and pintle.		
		Receive spare parts roll and spare bolt assembly from loader.		Hand spare parts roll and spare bolt assembly to bow gunner.
	Take one box caliber .30 ammunition.			
Command: DISMOUNT. Dismount via right stowage box and fender.	Dismount to left stowage box. Procure tripod.			
Receive two boxes caliber .30 ammunition from loader.	Dismount. Set up tripod.	Pass bow machine gun to driver.	Dismount and receive bow gun from bow gunner.	Pass two boxes caliber .30 ammunition to tank commander.

Help mount machine gun. Serve gun as No. 1.	Procure carbine and six magazines caliber .30 carbine ammunition. Dismount with carbine, spare parts roll, and spare bolt assembly. Receive one box caliber .30 ammunition from loader.	Mount bow gun and serve gun as No. 2.	Pass one box caliber .30 ammunition to bow gunner. Take submachine gun and six magazines caliber .45 ammunition. Dismount via left fender and provide security for machine gun crew with submachine gun. (Under certain conditions the loader may remain in the turret, connect the breakaway plugs, and maintain contact with the platoon leader.)
Act as squad leader of machine gun crew.	Serve gun as No. 3.		

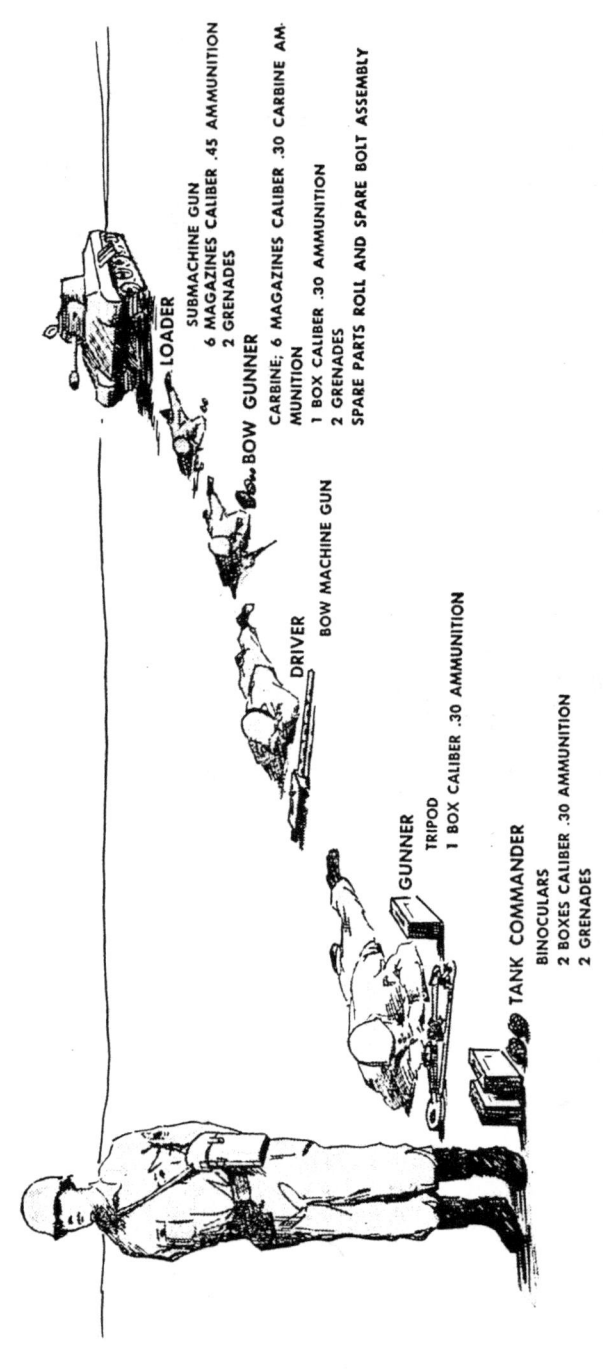

Figure 6. Dismounted action, crew formed for drill (tank commander instructing).

46

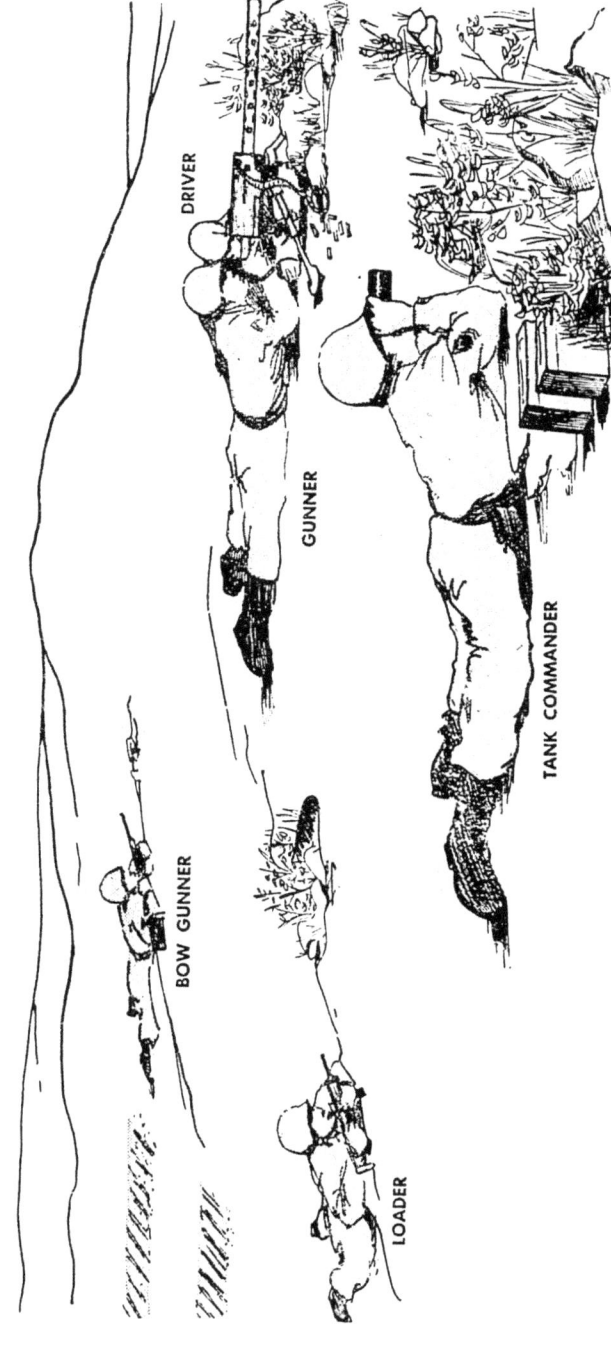

Figure 7. Posts of dismounted crew in action.

b. The dismounted crew moves to the position indicated by the tank commander, which, during drill, usually is 5 yards in front of the tank. The crew members take the posts and perform the duties prescribed for gun drill in FM 23-55.

c. In combat, tank personnel ordinarily will *not* fight on foot. Only in emergencies when necessary to defend a disabled tank or evacuate a destroyed tank, should tank crewmen fight outside their vehicle. However, since tank personnel are employed often to furnish close-in security for their own vehicles, they must be thoroughly familiar with the use of weapons which can be dismounted from the tank.

42. TO REMOUNT AFTER DISMOUNTED ACTION.

Tank commander	*Gunner*	*Bow gunner*	*Driver*	*Loader*
Command: OUT OF ACTION. Supervise taking gun out of action	Help dismount machine gun; fold tripod. Stow tripod. Mount to turret with box caliber .30 ammunition.	Take mounted post (leave caliber .30 ammunition in front of tank). Stow carbine and magazines.	Dismount machine gun.	Take mounted post. Stow submachine gun and magazines.

Pass caliber .30 ammunition to loader.	Enter turret and take mounted post.	Receive and mount bow machine gun. Stow ground accessories. Pass spare parts roll and spare bolt assembly to loader.	Pass machine gun to bow gunner.	Receive remaining caliber .30 ammunition from tank commander and stow.
Take mounted post.			Take mounted post.	Receive spare parts roll and spare bolt assembly from bow gunner and stow.
Stow grenades.		Return grenades.		Stow grenades.
Connect breakaway plugs.	Connect breakaway plugs.	Connect breakaway plugs.	Connect breakaway plugs.	Connect breakaway plugs.
Command: REPORT.	Report "Gunner ready."	Report "Bog ready."	Report "Driver ready."	Report "Loader ready."

43. TO ABANDON TANK. If it becomes necessary to abandon tank, the crew proceeds as in paragraph 16 or 17, with the following changes or additions:

a. If time permits deliberate action, the tank commander displays the flag signal: DISREGARD MY MOVEMENTS, and supervises the disabling of those weapons remaining in the tank. Back plates are removed from machine guns and the firing pin and guide from the tank gun (howitzer). Like items in spare parts kit are also removed. Individual weapons and maximum ammunition loads are carried.

b. Ordinarily the tank is abandoned as a result of a direct hit which either causes it to catch fire or disables it so that it becomes a vulnerable target. At the command ABANDON TANK, crew members open the hatches, climb out, and take cover at a safe distance from the tank. In case of fire, it is particularly important to hold the breath until clear of the vehicle. Inhaling fumes and smoke may cause serious injury.

44. TO DESTROY TANK. When the command DESTROY TANK is given, crew members first remove equipment that is to be carried away. They then destroy the tank, and remaining weapons, ammunition, and equipment, as prescribed in section XI.

45. ACTION IN CASE OF FIRE. a. Fire in engine compartment.

The first crew member to discover the fire calls "Engine fire."

Tank commander	*Gunner*	*Bow gunner*	*Driver*	*Loader*
Disconnect breakaway plugs.	Disconnect breakaway plugs.	Disconnect breakaway plugs.	Disconnect breakaway plugs.	Disconnect breakaway plugs.
Dismount to rear deck.		Operate fixed extinguisher control.	Shut off engine.	Dismount to rear deck.
Receive hand fire extinguishers.	Traverse turret 90° right or left.	Dismount.	Dismount with hand extinguisher; carry to rear of tank.	Open engine compartment grilles nearest fire.
Prepare to use hand extinguisher if fire has not been put out.	Pass hand fire extinguisher to tank commander.	Go to rear of tank.	Pass hand extinguisher to tank commander if needed.	Help tank commander as directed.
	Dismount.			
	Stand by to help where needed.	Stand by to help as needed.		

b. Fire in fighting compartment. The first crew member to discover the fire calls: "Turret (hull) fire." The tank is stopped and the engine shut off. Fire extinguishers are passed to the men nearest the fire, and nearby crew members help in any way possible to extinguish the fire. The turret is traversed when necessary. The tank commander supervises the work and orders the crew to dismount if the fire gets beyond control.

46. CONDUCT OF INSTRUCTION IN DISMOUNTED ACTION. a. Long, intensive drill is essential to the execution of effective dismounted action. Satisfactory results can be obtained in drill only by painstaking repetition of each movement. The technique of mounting and dismounting of all crew members is observed carefully by tank and platoon commanders and altered, when necessary, before habits are formed. Once each man has learned the most efficient method of mounting and dismounting, he is encouraged to adhere rigidly to it.

b. Training in dismounted action is undertaken best in the field, rather than in the tank park. Crews must be required to dismount to fight on foot on all types of terrain and under various simulated combat conditions, with full loads of ammunition. Rough terrain complicates the problem of dismounting through the escape hatch, and develops the ingenuity and physical agility to an extent not possible in tank park training.

c. Instructors must explain and demonstrate to tank crews that a high degree of proficiency in dismounted action is necessary to their safety and success in combat. They must point out that dismounted action from a disabled tank under small-arms fire usually is practicable only through the escape hatches; therefore, skill in use of the escape hatches is essential. The crew keeps the escape doors clean and well lubricated so that they can be released positively and without delay. Frequent inspection of the mechanisms is made by the tank commander to see that they are in serviceable condition, and that the gaskets are in place and sealed against water and air.

47. MOUNTING AND DISMOUNTING PRECAUTIONS.
a. Crew members mount and dismount at the front of the tank except during combat or range practice when they may mount in the most expeditious manner.

b. In mounting and dismounting, weapons and sighting equipment should not be used as aids, for example—

(1) Stepping on the gun (howitzer) barrel, gun shield, periscope covers, or machine guns.

(2) Using the tube, gun shield, or machine guns as supports.

(3) Using the shoulder guard as a step in entering or leaving the turret.

48. SAFETY PRECAUTIONS FOR FIRE PREVENTION.
a. Smoking in or on the tank is prohibited.

b. During fueling, a crew member stands on the rear deck holding a fire extinguisher ready for immediate use with the nozzle trained on the fuel inlet.

c. Use of gasoline for cleaning any part of the tank is prohibited.

49. SAFETY PRECAUTIONS IN OPERATING TANK. a. Crash helmets or helmet liners are worn at all times inside the tank.

b. In operating cross-country, the tank commander warns the driver and crew when the tank is approaching rough terrain.

c. Where practicable, the driver detours around rough or uneven ground to prevent damage to the tank or injury to its crew.

d. In traveling with hatches open over rough ground or through woods, crew members constantly check the engagement of the cover latching mechanisms and the security of covers in the open position.

e. The antenna is lowered to prevent contact with low branches and low-hanging wires, especially where the latter may carry high-voltage electricity.

f. The tank is driven in low range when being moved forward in confined spaces.

50. PARK AND BIVOUAC SAFETY PRECAUTIONS. a. Sleeping underneath, behind, and in front of tanks should be prohibited.

b. In moving a tank in park or bivouac, a guide is always employed to direct the movement. The driver will move the tank only when he has contact by sight or signal with the guide.

  (1) The guide's position is at least 10 feet in front of the tank and to one side, clear of its path, in directing the tank forward or backward.
  (2) At night, the guide is charged especially with seeing that the path ahead of and behind the tank is clear of personnel, particularly those sleeping on the ground.

(3) The guide moves at a walk to avoid stumbling on uneven ground.

51. MISCELLANEOUS SAFETY PRECAUTIONS.

a. Prior to commencing tank firing and prior to reporting tank weapons as cleared after firing, an inspection is made of bores, chambers, and T-slots to insure clearance. After clearance, T-blocks are inserted in the receivers of machine guns.

b. Tank weapons, except the antiaircraft gun, are fired only when the driver's and bow gunner's hatches are closed.

c. Before turning on the power traverse, a check is made to see that the driver and bow gunner are clear of the gun (howitzer).

d. Care will be taken, while working about a running engine, to keep fingers and hands away from fans, fan belts, drive shafts, other moving parts, and exhaust manifolds and exhaust pipes.

e. Ammunition will be stowed securely.

f. Ammunition and gasoline will not be transported on the rear deck.

g. Items of equipment will not be carried on the rear deck in a manner to block the air inlet and exit grilles.

h. There is danger of monoxide poisoning for the crew of a towed tank when a medium tank or a tank recovery vehicle is used as the towing vehicle. This danger is greatest when using a short hitch; such as the towing bar. Everyone except the driver should be kept out of the towed tank. The driver of the towed vehicle should be relieved frequently and kept under continuous observation from the towing vehicle.

SECTION VIII

EVACUATION OF WOUNDED FROM TANKS

52. GENERAL. Wounded members of the tank crew normally will evacuate themselves from disabled tanks or be removed by their fellow crew members. The utmost speed is necessary in order to save the lives of those who are unhurt as well as the life of the casualty. A tank on fire can trap its crew in a matter of seconds, and an enemy who has determined the range and has disabled a tank probably will continue shooting until the vehicle burns. It is essential, therefore, that all crew members become extremely proficient in utilizing the quickest methods of removing one another from the tank. Speed is the primary requisite; care in handling will be stressed only where it has been possible to move the tank to cover. If the action has ceased momentarily, or the tank has been able to disengage itself without hindering the accomplishment of the mission, the casualty is removed on the spot and carried to a protected place, where emergency first aid is administered. Otherwise, the action will be continued until an opportunity is presented to remove casualties.

53. METHODS OF EVACUATION. Methods of evacuation described herein are based on the employment of a two-man team, the largest number that can work effectively around a single hatch opening. In some cases, a third man will be able to give considerable help from inside by placing belts around the wounded man or by moving him to a position where he can be grasped from above. Necessity for swift action usually will require that the casualty be grasped for removal by portions of his clothing or by the arms. If an arm is broken, or if there are other injuries which will be aggravated by these procedures and, if time allows, some form of sling may be improvised which will protect the part from further injury. Any equipment which is available immediately, such as pistol belts, web belts, or field bag straps, may be used for this purpose.

54. EVACUATION DRILL, GENERAL. a. This paragraph contains general information which may be used as a guide in practicing the evacuation of crew members from any position. Paragraph 55 and 56 describe drills in evacuating personnel from driving compartment and turret, respectively. During drill, composition of the evacuating team should be changed frequently to provide practice for all members of the crew in meeting various emergencies.

b. The first member of the crew to discover that another is badly wounded calls: "Bog (loader) (tank commander) wounded." If the tank is not engaged actively and the tank commander decides that evacuation is necessary, he commands: EVACUATE BOG (LOADER) (TANK COMMANDER). Crew members dismount, unless one man is needed to help from inside, and the two nearest the hatch above the wounded man (Nos. 1 and 2 in pars. 55 and 56), take stations at that hatch to act as the evacuation crew. If the man nearest the casualty, in the tank, sees that his help is needed, he stays inside and immediately begins to arrange a sling, or take whatever steps he can to speed the operation. The remaining crew member removes the first-aid kit from its armored box on the outside of the vehicle and helps in lowering the casualty to the ground. Before leaving the wounded man, first aid is administered, he is moved to a sheltered position, which is marked so that he will not be run over, the tank commander reports by radio that he has lost one or more men and gives the location where they may be found.

55. PROCEDURE TO EVACUATE CASUALTY FROM DRIVING COMPARTMENT (fig. 8).

Tank commander commands: EVACUATE BOG. Driver or loader unlocks bow gunner's hatch from the inside. No. 2 opens the hatch from the outside.

No. 1	*No. 2*
Take position on edge of hatch.	Take position on edge of hatch.
Reach into hatch and grasp hands of casualty, straightening him in seat if necessary.	
Cross arms over chest.	Grasp nearest hand of casualty when his arms are crossed.
Raise and rotate casualty so that he faces to the rear.	Raise casualty and help rotate him.
Seat casualty on front rim of hatch; support him in this position while No. 2 jumps to ground.	Help seat casualty; jump to ground, and go to front of tank.
Lower trunk of casualty into arms of No. 2.	Receive and support trunk of wounded man, holding him beneath arms, around chest.

① Medium (M26) tank crew evacuating bow gunner.

② Medium (M26) tank crew evacuating bow gunner down front slope plate.

Figure 8. Evacuation of wounded.

No. 1	*No. 2*
Lift legs out of hatch as No. 2 lowers casualty along slope plate.	Lower casualty along slope plate and support him until No. 1 can reach ground and assist.
Jump to ground; help No. 2 place casualty in carry position.	Place casualty in carry position.
Carry casualty to protected area.	Help No. 1 carry casualty to protected area.

56. PROCEDURE TO EVACUATE CASUALTY FROM TURRET.
Tank commander commands: EVACUATE LOADER, and dismounts to rear deck to act as No. 1. The gunner, as No. 2, stays in the turret to lift casualty from below. He unlocks loader's hatch and opens it with the help of No. 1.

No. 1	*No. 2*
Take position on turret beside loader's hatch.	Raise casualty as high as possible in hatch opening, holding him around chest.
Grasp casualty under arms.	
Raise casualty through hatch, and seat on edge.	Help No. 1 raise casualty by lifting from below.
Hold casualty while No. 2 dismounts to stowage box.	Dismount to stowage box.
Dismount to stowage box.	Support casualty while No. 1 dismounts to stowage box.
Lift casualty from turret; lay along stowage box.	Help No. 1 lift casualty and lay at edge of tank.
Jump to ground.	Jump to ground.
Lift trunk of casualty off tank.	Lift casualty's legs and feet off tank.
Carry casualty to protected area.	Help carry casualty to protected area.

SECTION IX

INSPECTIONS AND MAINTENANCE

57. GENERAL. a. The tank commander is responsible for insuring that required inspections are made. Mechanical efficiency is essential to tank unit operation; therefore, each tank must be inspected systematically at intervals during each day of use. Defects then can be discovered and corrected before they result in serious mechanical damage or failure. Crew members make their individual inspections and report the results to the tank commander, who, in his own report, lists all items requiring the services of maintenance personnel. In supervising driver maintenance or other services performed at periodic intervals and from day to day, the tank commander delegates responsibility to crew members as necessary. Maintenance procedure omitted from this manual is set forth in detail in TM 9-735.

b. Inspections are made of all personal equipment and weapons, communication equipment, vehicle equipment, and mechanical features of the vehicle. In combat, a check is made to determine whether crew members have applied properly protective cream to minimize the effects of flash burns. Inspections of instruments, lights, siren, tracks, suspension system, and engine performance are made in accordance with provisions of appropriate technical manuals. The driver fills in NME Form 110 (Vehicle and Equipment Operational Record), indicating thereon deficiencies or maintenance work required. He must prepare this form carefully. Any irregularity entered on NME Form 110 and not repaired before the tank is again used will be reentered on the same form continually until the deficiency has been corrected.

c. In succeeding paragraphs (58–62), the duties of crew members in performing tank inspections are tabulated in chart form to be used as an aid in training. However, the steps listed on these charts need not be followed exactly—though they should be used as a guide—in the later conduct of crew drills.

58. BEFORE-OPERATION INSPECTION. (Tank-locked and covered by tarpaulin.) For training purposes, the inspection is divided into three phases. Each phase is completed before beginning the next. Crew members coordinate their respective operations to make the best use of the time available. They procure tools as needed, and report and correct deficiencies as found. The turret is traversed as necessary to facilitate the various operations.

PHASE A

Tank commander	*Gunner*	*Bow gunner*	*Driver*	*Loader*
Command: FALL IN: PREPARE FOR INSPECTION.	Stand inspection.	Stand inspection.	Stand inspection.	Stand inspection.
Inspect crew. Command: PERFORM BEFORE-OPERATION INSPECTION.	Help remove tarpaulin.	Inspect ground beneath tank for fuel, oil, or water leaks.	Remove and fold tarpaulin.	Help remove and fold tarpaulin.
Supervise inspection made by other crew members and filling out of NME Form 110 by driver.				

63

PHASE A—Continued

Tank commander	Gunner	Bow gunner	Driver	Loader
	Mount right fender; unlock loader's hatch; enter tank.		Lay tarpaulin to left of tank.	Check contents of stowage boxes.
Inspect tracks and tank suspension. (Visual check is adequate for daily inspection of track pin nuts.)		Mount to rear deck via front of tank.	Begin filling out NME Form 110; fill out during inspection. Procure hand tools; lay out on tarpaulin and check.	Remove and stow muzzle covers.
	Unlock vision cupola. Remove breech covers; clear turret guns.	Check engine fuel level. Open engine compartment grilles.		Unlock gun traveling rest. Check outside equipment.

Help bow gunner open engine grilles.	Check oil levels; differential; tank engine; auxiliary engine.
Check rolling and stowage of camouflage net.[1]	Move into bow gunner's compartment; unlock and open bow gunner's hatch.
	Remove breech cover and clear bow gun.
	Pass covers to loader.
	Move into turret.
	Unlock driver's hatch.
Receive and stow breech covers.	Take mounted post.

[1] The camouflage net is folded into a 12-foot roll for stowage as follows: Fold opposite sides toward center, leaving 3 feet between edges. Lap one fold over the other with folded edges touching to make section 12' by 45'. Fold two ends toward center and double over. Repeat procedure in same direction. Roll net and stow with loose edge toward tank so that it will not catch on branches, etc.

PHASE A—Continued

Tank commander	Gunner	Bow gunner	Driver	Loader
Command: REPORT.	Report "Gunner ready."	Report "Bog ready."	Report "Driver ready."	Report "Loader ready."

PHASE B

Tank commander	Gunner	Bow gunner	Driver	Loader
Command: PERFORM PHASE B. Procure cleaning rods; swab bores of cannon and all machine guns. Help gunner make sight adjustment.	Traverse turret manually one revolution to the left; check azimuth indicator.[2] Make sight adjustment.		Open hatch; take mounted post. Close battery master switch. Open left fuel valve.	Take mounted post. Check the following: Rations; 90-mm (105-mm) ammunition;
			Check the following: Engine coolant level; auxiliary engine air cleaner;	

Stow cleaning rods. Check all hatch covers (note condition of gaskets).

Check the following: Power traverse mechanism and hydraulic oil; elevating mechanism; firing controls.

breathers for transmission, differentials, and final drives; auxiliary engine crankcase; fan and generator belts.

Check the following: Steering levers and linkage; range selector lever (place in neutral); parking brake; fuel cut-off operation; instruments; siren; service lights and black-out lights.

lubrication guide; tank engine oil filter; crankcase breather; engine compartment drain valves; priming pump operation; auxiliary engine operation (warn bow gunner when starting); machine gun tools and spare parts; flare launcher, flares, and cartridges;

---

[2] Traverse is made piecemeal and even may be reversed for short distances to coordinate with and facilitate other operations and checks.

PHASE B—Continued

Tank commander	Gunner	Bow gunner	Driver	Loader
Help driver check lights.	Level howitzer (M45 only). Check gyrostabilizer (oil level, connections, cleanliness, and operation). Turn off gyrostabilizer. Engage manual elevation mechanism. Help loader check air cleaners.	Listen for operation of fuel cutoff. Check: Engine accessories for security and adjustment; engine compartment for fuel, oil, or water leaks.		oilcan; spare hydraulic, recoil, and engine oil. Check: Air cleaners; batteries; hull drain valve beneath turret.

Command				
Command: REPORT.	Report "Gunner ready."	Report "Bog ready."	Report "Driver ready."	Report "Loader ready."

Command: PERFORM PHASE C.

Check sprocket ring cap screws.	Help loader attach shell bag. Check the following: gun (howitzer), tools, and spare parts; hand fire extinguisher;	*PHASE C*[3] Observe condition of exhaust. Check engine for leaks, vibrating accessories, or parts. With engine at operating tem-	Start engine (operate at 1,100 rpm until temperature reaches 100° F.). During warm up, check: Instruments;	Attach empty shell bag. Mount AA gun; check gun and mount; adjust headspace. Check the following:

---

[3] The flame thrower, on tanks so equipped, is checked in this phase. The crew member using the weapon checks its condition, its mechanism, and the fuel level in its tank in accordance with the appropriate published guide. Where it is an alternate weapon to the bow machine gun, it is mounted on order of the tank commander.

PHASE C—Continued

Tank commander	Gunner	Bow gunner	Driver	Loader
	periscope, spare, and spare heads (including knob settings); gunner's quadrant and case; elevation quadrant; telescope and mount; gun (howitzer) and mount.	perature, check transmission oil level. Close engine compartment grilles.	engine for smoothness of operation and ususual noises; magnetos.	Coaxial gun and mount; adjust headspace; hand grenades; AA and coaxial machine gun ammunition; caliber .45 ammunition; periscope, spare, and spare heads.
Help bow gunner close engine grilles. Direct driver to move tank forward two tank lengths.		Dismount.		
Walk beside right idler.	Clean chamber and breech mechanism. Check recoil oil.	Walk beside left idler.	Drive tank forward at slow speed two tank lengths.	

Inspect that part of track not previously visible; observe action and condition of support rollers, track, and tank suspension as tank moves forward.	Lock gun (howitzer) and turret in traveling position.[4]	Inspect that part of track not previously visible; observe action and condition of support rollers, track, and tank suspension as tank moves forward.
	Connect breakaway plugs.	
Direct tank to rear.	Place tools in bag and stow.	Drive tank to rear as directed. Check the following: fixed fire extinguishers and control; hand fire extinguisher;
Roll and stow tarpaulin.	Help tank commander with tarpaulin.	
Take mounted post.	Take mounted post.	Connect breakaway plugs.

[4] If turret is to be locked to rear, gun must be secured in gun traveling rest; turret traversing lock will not be engaged at the same time.

PHASE C—Continued

Tank commander	Gunner	Bow gunner	Driver	Loader
Check the following. Carbine and ammunition; periscope, spare, spare heads, and periscope holder; spare vision blocks; four grenades; compass; flag set; binoculars.		Open right fuel valve. Check the following: Bow gunner's driving controls; bow gun, mount, and ammunition—adjust headspace;	Check: Submachine gun and ammunition; periscope, spare, and spare heads; escape hatch.	

Mount antenna.	caliber .30 ammunition; first-aid kit;[5] escape hatch; forward hull drain valve; periscope, spare, and spare heads.			
Make radio check (par. 9).				
Connect breakaway plugs.	Connect breakaway plugs.	Connect breakaway plugs.	Connect breakaway plugs.	
Command: REPORT (interphone check).				
	Report "Gunner ready."	Report "Bog ready."	Report "Driver ready."	Report "Loader ready."
Report "Ready" to platoon leader.				

[5] First-aid kit is carried in its bracket behind the bow gunner until tank reaches the assembly area when it is transferred to its armored container at the rear of the left fender.

59. INSPECTION DURING OPERATION.
This is a continuous process for all crew members.

Tank commander	*Gunner*	*Bow gunner*	*Driver*	*Loader*
Remain alert to unusual noises or conditions.	Check operation of: elevating and traversing mechanisms; gyrostabilizer, (M45 only).	Watch instruments.	Check all instruments carefully.	Check stowage of equipment in turret.
Check radio and interphone system.		Listen for unusual noises.	Check controls.	
Check security of: Radio antenna; outside fixtures and equipment.	Check security of: turret lock and gun (howitzer) traveling lock.	Check security of bow gun.	Listen for unusual noises.	Check security of: Coaxial gun; radio; antiaircraft gun.

60. INSPECTION AT THE HALT.
The length of halt is the basis for determining how much of the following inspection will be completed and the priority to be given various operations. The tank commander will be informed of the length of halt and will indicate how much time is to be allotted to inspection and how much for relief of the crew members.

Tank commander	Gunner	Bow gunner	Driver	Loader
Command: PERFORM HALT INSPECTION.				
Disconnect breakaway plugs.	Disconnect breakaway plugs.	Disconnect breakaway plugs.	Disconnect breakaway plugs.	Disconnect breakaway plugs.
Turn on radio speaker.	Release turret lock.[6]	Dismount to rear deck.	Idle engine (run 5 minutes before stopping). Check instruments.	Man AA gun.
Dismount to rear deck.	Check hand traverse.			
Help bow gunner open engine grilles.	Check sight adjustment.	Open engine compartment grilles.	Dismount to rear deck.	
Clean all turret periscopes, telescope, and vision blocks.	Check the following: coaxial gun and mount; gun (howitzer) and mount;	Check transmission oil level.	Check engine operation.	
	Check: Firing controls; power traverse; auxiliary engine operation (leave on if	Check engine fuel and coolant levels.	Inspect engine compartment.	
Supervise halt inspection.		Check tank and auxiliary engine	Take mounted post.	

[6] On nontactical marches 90-mm. gun traveling rest will have to be unlocked and locked by bow gunner at rear of tank.

Tank commander	*Gunner*	*Bow gunner*	*Driver*	*Loader*
Inspect tracks and tank suspension.	necessary); air cleaners.	and differential oil levels.	Stop engine (use fuel cut-off).	
Help bow gunner close grilles.		Close engine compartment grilles.	Check driver's and bow gunner's compartments for oil leaks.	
	Lock gun (howitzer) in traveling position.[6]	Dismount. Check hubs of sprockets, wheels, and rollers for leaks and excessive temperatures.	Check the following: steering levers and linkage; range selector lever.	
	Lock turret lock.			
	Check stowage of equipment in turret.	Check outside equipment.	Parking brake.	
	Connect breakaway plugs.	Help driver check lights.	Service and blackout lights.	
		Check under tank for fuel, oil, or water leaks.	Clean periscopes. Connect breakaway plugs.	

Take mounted post.	Check towing shackles. Take mounted post. Clean periscopes.		
Turn off radio speaker. Connect breakaway plugs. Command: REPORT		Connect breakaway plugs.	Connect breakaway plugs.
	Report "Gunner ready."	Report "Bog ready."	
		Report "Driver ready."	Report "Loader ready."

61. AFTER-OPERATION MAINTENANCE. *a.* Immediately after operation, the tank is given servicing and maintenance needed to prepare it in every way for sustained action. This servicing covers all the points listed in the before-operation inspection and covers them in the same order. However, in a number of cases, procedures are reversed. For example, the tank is locked at the end of the inspection instead of being unlocked as it is at the beginning; the check for leaks under the tank is more effective after it has stood for a while; battery switches are turned off rather than on and only after all checks have been made requiring use of battery power; and equipment is covered and stowed rather than being uncovered and made ready for use.

b. The tank is cleaned, serviced, and replenished (fuel and oil (all types), grease, coolant, ammunition (all types), first-aid kit, water, and rations). *All precautions against fire will be observed while refueling.* Crew members perform the following additional operations not covered in the before-operation inspection:

Tank commander	Gunner	Bow gunner	Driver	Loader
Command: PERFORM AFTER - OPERATION MAINTENANCE.	Clean all weapons.		Idle engine five minutes before stopping.	Help gunner clean weapons.
Complete and forward NME Form 110 to platoon leader, together with report of maintenance, fuel, lubricants, ammunition, and rations required.		Help driver clean tank.	Clean tank suspension and outside of tank.	
		Help gunner clean weapons.		

62. PERIODIC ADDITIONAL SERVICES. These services are performed weekly in garrison; after each field operation in combat and on maneuvers.

Tank commander	Gunner	Bow gunner	Driver	Loader
Command: FALL IN; PREPARE FOR INSPECTION.				
Inspect crew.	Stand inspection.	Stand inspection.	Stand inspection.	Stand inspection.
Command: PERFORM PERIODIC INSPECTION.				
Supervise inspection.	Mount to turret. Clean turret. Clean and paint any rust spots in turret.	Mount to rear deck; help driver clean engine and engine compartment.	Mount to rear deck; open grilles; clean engine and engine compartment.	Mount to turret. Clean batteries and case.
				Test batteries with hydrometer.
	Dismount.	Take mounted post.	Take mounted post.	Bring cells to proper water level.
		Clean bow gunner's compartment and right interior of hull.	Clean driver's compartment and left interior of hull.	Operate auxiliary generator to charge batteries.

Tank commander	Gunner	Bow gunner	Driver	Loader
		Operate and check hull drain valve in bow gunner's compartment.		Operate and check drain valves in engine compartment and beneath turret.
		Clean and paint any rust spots in bow gunner's compartment.		
	Tighten all track pin nuts and inspect track.	Dismount.	Drive tank forward as required for tightening track pin nuts.	Dismount.
				Help gunner tighten track pin nuts and inspect track.
	Help perform 250-mile lubrication.	Help perform 250-mile lubrication.	Dismount. Perform 250-mile lubrication, referring to appropriate guide.	Help perform 250-mile lubrication.
		Help driver close engine grilles.	Close engine compartment grilles.	

| Command: REPORT. | Report "Gunner ready." | Report "Bow ready." | 'Take mounted post; clean and paint any rust spots in driver's compartment. | Report "Driver ready." | Report "Loader ready." |

SECTION X

SIGHT ADJUSTMENT

63. FIELD CHECK OF SIGHTS. Frequent checking of the sights is vital in the field. A rapid check may be performed as follows: Select, as an aiming point, an object at least 1,500 yards away, preferably one having distinct straight lines which intersect. By moving the tube, place the intersection of the bore sight cross of the *telescope* on the aiming point. Check to see whether the intersection of the bore sight cross of the *periscope* is on the same point. If the two bore sight crosses are in agreement, it may be assumed safely that the sights still are in adjustment. If they are not, the adjustment has slipped, and a new sight adjustment must be made for both sights, using the normal boresighting method. At the same time, the vane sight should be checked for alinement. This field check is not a positive proof of proper sight adjustment but, for practical purposes, it can be assumed that if the sight pictures of the telescope sight and the telescope in the periscope coincide, the sights are in proper adjustment. If time permits, the sights should be checked and adjusted exactly as in field boresighting. Complete coverage of sight alinement by boresighting may be found in **FM 23-100**.

SECTION XI

DESTRUCTION OF EQUIPMENT

64. GENERAL. a. The destruction of matériel is a command decision to be carried out only on authority delegated by the division or higher commander. This usually is made a matter of standing operating procedure. *It is ordered only after every possible measure for preservation or salvage of the matériel has been taken and, when in the judgment of the military commander concerned, such action is necessary to prevent—*
  (1) Its capture intact by the enemy.
  (2) Its use by the enemy, if captured, against our own or allied troops.
  (3) Its abandonment in the combat zone.
  (4) Knowledge of its existence, functioning, or exact specifications from reaching enemy intelligence agencies.

b. The principles followed are—
  (1) Methods for the destruction of matériel subject to capture or abandonment in the combat zone must be adequate, uniform, and easily followed in the field.
  (2) Destruction is as complete as possible within limitations of time, equipment, and personnel. If thorough destruction cannot be completed, the most important features of the matériel are destroyed, and parts which cannot be duplicated easily and are essential to the operation or use of the matériel are ruined or destroyed. *The same essential parts are destroyed on all like units to pre-*

vent the enemy's constructing one complete unit from several damaged ones.

c. Crews are trained in employing prescribed methods of destruction. *Training does not involve actual destruction of matériel.*

d. Certain methods of destruction require special tools and equipment, such as TNT and incendiary grenades, which may not be items of issue. The issue of such special tools and material, the vehicles for which issued, and the conditions under which destruction will be effected are command decisions, and depend upon the tactical situation.

e. For methods of destruction of small arms see the field manual pertaining to the weapon.

f. The methods below are given in order of effectiveness. If method No. 1 cannot be used, destruction is accomplished using other methods in the priority shown. Adhere to the sequence in which operations are performed.

65. DESTRUCTION OF THE TANK GUN (HOWITZER).

Remove the periscopic and telescopic sights. *If evacuation is possible, carry the sights.* If evacuation is not possible, thoroughly smash the sights; also the spare sights.

a. Method No. 1.

(1) Open drain plug on recoil mechanism, allowing recoil fluid to drain. It is *not* necessary to wait for the recoil fluid to drain completely before firing the cannon as described in (4) below.

(2) Place an armed (safety pin removed) antitank grenade, HE round, or armed (safety pin removed) antitank rocket in the tube

with the ogive nose end toward and about 6 inches in front of the HE shell described in (3) below.

(3) Set the fuze on an HE shell at SQ (superquick), insert the shell in the gun (howitzer), and close the breech. An armor-piercing shell cannot be used.

(4) Attach a strong cord (if nothing better, improvise a lanyard with tent ropes) to the firing linkage so that the gun (howitzer) can be fired by pulling the cord. Dismount from the tank, take cover down and to the rear, and fire the piece. Elapsed time: approximately 2 to 3 minutes.

b. Method No. 2. Insert three to five TNT blocks in the bore near the muzzle, eight to ten in the chamber of the gun (howitzer). Close the breechblock as far as possible without damaging the safety fuze. Plug the muzzle tightly with earth to a distance of approximately 12 inches from the muzzle. Detonate the TNT charges simultaneously.

c. Method No. 3. With another gun (howitzer), fire HE or AP projectiles at the tube of the gun (howitzer) until it is rendered useless.

d. Method No. 4. Insert four unfuzed incendiary grenades, end to end, midway in the tube at 0° elevation. Ignite the four grenades with a fifth equipped with a 15-second delay detonator. The metal from the grenades will fuse with the tube and fill the grooves. Elapsed time: 2 to 3 minutes.

66. DESTRUCTION OF THE GYRO-STABILIZER.

a. Drain the oil from the system.

b. Smash the oil lines.

c. Smash the control box.

d. Place an incendiary grenade on the control box and pull the pin.

67. DESTRUCTION OF MACHINE GUNS. a. Method No. 1.

(1) Caliber .30 machine gun. Field strip. Use the barrel as a sledge. Raise the cover until vertical and smash it down to the front. Deform and break the back plate, and deform the **T**-slot. Wedge the lock frame, back down, into the top of the receiver between the top plate and the extractor cam; place the chamber end of the barrel over the lock frame depressors and break them off. Insert the barrel extension into the back of the receiver, allowing the shank to protrude; knock off the shank by striking with the barrel from the side. Deform and crack the receiver by striking it with the barrel at the side plate corners nearest the feedway. Elapsed time: 2½ minutes.

(2) Caliber .50 machine gun. Field strip. Use the barrel as a sledge. Raise the cover; lay the bolt in the feedway; lower the cover on the bolt and smash the cover down over the bolt. Deform the back plate. Wedge the buffer into the rear of the Receiver allowing the depressors to protrude; break off the depressors by striking them with the barrel. Lay the barrel extension on its side. Hold it down with one foot; break off the shank. Deform the receiver by striking the side plates just back of the feedway. Elapsed time; 3½ minutes.

b. Method No. 2. Insert the bullet point of a complete round into the muzzle and bend the case slightly, distending the mouth of the case to permit removing the bullet. Spill powder from the case, retaining sufficient powder to cover the bottom of the case to a depth of approximately one-eighth inch. Reinsert the pulled bullet, point first, into the case mouth. Chamber and fire this round with the reduced charge; the bullet will stick in the bore. Chamber one complete round; lay the weapon on the ground, and fire with a 30-foot lanyard. Use the best available cover; otherwise, this means of destruction is dangerous to the person destroying the weapon. Elapsed time; 2 to 3 minutes.

c. Machine gun tripod mount, caliber .30, M2. Use the machine gun barrel as a sledge. Deform the traversing dial. Fold the rear legs; turn the mount over on its head; stand on folded rear legs; knock off the traversing dial locking screw and pintle lock, and deform the head assembly. Deform the folded rear legs so as to prevent unfolding. Extend the elevating screw and bend it by striking it with the barrel; bend the pintle yoke. Elapsed time: 2 minutes.

68. DESTRUCTION OF TANK. a. Method No. 1.

(1) Remove and empty the portable fire extinguishers. Smash the radio (par. 71). Puncture the fuel tanks. Use fire of caliber .50 machine gun, or a cannon, or use a fragmentation grenade for this purpose. Place TNT charges as follows: 2 pounds on each side of the transmission between transmission and engine oil cooler; 3 pounds between the carburetors in the **V** of the engine

blocks. Insert tetryl nonelectric caps with at least 5 feet of safety fuze in each charge. Ignite the fuzes and take cover. Elapsed time: 1 to 2 minutes (if charges are prepared beforehand and carried in the vehicle).

(2) If sufficient time and materials are available, additional destruction of track-laying vehicles may be accomplished by placing a 2-pound TNT charge at the center of each track-laying assembly. Detonate charges as described in (1) above.

(3) If charges are prepared beforehand and carried in the vehicle, keep the caps and fuzes separated from the charges until used.

b. Method No. 2. Remove and empty the portable fire extinguishers. Smash the radio (par 71). Puncture the fuel tanks (see *a* (1) above). Fire on the vehicle using adjacent tanks, antitank guns, or other artillery, or antitank rockets or grenades. Aim at the engine, suspension, and armament in the order named. If a good fire is started, the vehicle may be considered destroyed. Elapsed time: about 5 minutes per vehicle. Destroy the last remaining vehicle by the best means available. Be certain that destruction is complete to prevent the enemy from obtaining usable parts.

69. DESTRUCTION OF AMMUNITION. a. General. Time usually will not permit the destruction of all ammunition in forward combat zones. When sufficient time and materials are available, ammunition is destroyed as indicated below. At least 30 to 60 minutes are required to destroy adequately the ammunition carried by combat units. In general, the methods and safety precautions outlined in TM 9–1901 are followed whenever possible.

b. Unpacked complete round ammunition.

(1) Stack the ammunition in small piles. (Small-arms ammunition may be heaped.) Stack or pile most of the available gasoline in cans and drums around the ammunition. Throw onto the pile all available inflammable material, such as rags, scrap wood, and brush. Pour the remaining gasoline over the pile. Sufficient inflammable material is used to insure a very hot fire. Ignite the gasoline and take cover.

(2) Cannon ammunition is destroyed by sympathetic detonation, using TNT. Stack the ammunition in two stacks, about 3 inches apart, fuzes toward each other. Use 1 pound of TNT to four or five rounds of ammunition. From cover, detonate all TNT charges simultaneously.

c. Packed complete round ammunition.

(1) Stack the boxed ammunition in small piles. Cover with all available inflammable materials, such as rags, scrap wood, brush, and gasoline in drums or cans. Pour other gasoline over the pile. Ignite and take cover. (Small-arms ammunition must be broken out of boxes or cartons before burning.)

(2) The destruction of packed complete round ammunition by use of TNT to obtain sympathetic detonation, is not practicable in forward combat zones. Satisfactory de-

struction involves putting TNT in alternate boxes of ammunition, a time-consuming job.

(3) In rear areas or fixed installations, sympathetic detonation may be used to destroy large ammunition supplies if destruction by burning is not feasible. Stack the boxes after placing in alternate boxes in each row, sufficient TNT blocks to insure the use of 1 pound per four or five rounds of cannon ammunition. Place the TNT blocks at the fuze end of the rounds. Detonate all TNT charges simultaneously. See FM 5-25 for details of demolition planning and procedure.

d. Miscellaneous. Grenades, antitank mines, and antitank rockets are destroyed by the method outlined for complete rounds in **b** and **c** above. The amount of TNT necessary to detonate these munitions is less than that required for detonating artillery shells.

70. FIRE CONTROL EQUIPMENT. All fire control equipment, including optical sights and binoculars, is difficult to replace. It is the last equipment to be destroyed. If personnel are evacuated, fire control equipment is carried insofar as possible. If evacuation is impossible, fire control equipment is destroyed thoroughly by smashing and burning.

71. DESTRUCTION OF RADIO EQUIPMENT. a. Books and papers. Instruction books, circuit and wiring diagrams, records of all kinds for radio equipment, code books, and registered documents are destroyed by burning.

b. Radio sets.
  (1) Shear off all panel knobs, dials, etc., with an axe. Break open the set compartment by smashing in the panel face; then knock off the top, bottom, and sides. The object is to destroy the panel and expose the chassis.
  (2) With the axe head, smash all tubes and circuit elements on top of the chassis. On the underside of the chassis, if it can be reached, use the axe to shear or tear off wires and small circuit units. Break sockets and cut unit and circuit wires. Smash or cut tubes, coils, crystal holders, microphones, earphones, and batteries. Break mast sections; break mast base at the insulator.
  (3) When possible, pile up smashed equipment, pour on gas or oil, and set it on fire. If other inflammable material, such as wood, is available, use it to increase the fire effect. Bury smashed parts whether burned or not.

SECTION XII

STOWAGE

72. GENERAL. a. The proper stowage of tank equipment is necessary for the efficient functioning of the tank and crew. First, each crew member must ascertain whether the equipment necessary to perform his duties is present. Second, and equally important, this equipment must be stowed in the proper place so as to be available when needed. In order that these conditions can be met, a list of vehicle stowage and special tools has been prepared for each vehicle. The lists for the Medium Tanks M26 and M45 are contained in appendixes II and III.

b. The list of vehicle stowage and special tools designates an exact location either on or within the tank for every piece of authorized equipment, including personal equipment.

73. STOWAGE OF PERSONAL WEAPONS. Personal weapons, with the exception of the bow gunner's carbine and the loader's submachine gun, normally will be worn by the individuals. The bow gunner's carbine and the loader's submachine gun will be stowed in the racks provided.

74. TRAINING IN STOWAGE. During training, tank crews should become familiar with the location of every piece of equipment on the tank. This can be achieved only through a continued series of drills stressing stowage and restowage of all vehicular and personal equipment. Crew commanders should rotate crewmen within the tank during these drills so that each man can learn thoroughly the proper and efficient stowage of his tank.

APPENDIX I

PREPARATION OF SUBJECT SCHEDULES

Given below is a sample subject schedule for use as an aid in planning and conducting training in the subjects covered by this manual. It is designed to provide for 31 hours of instruction, but can be expanded or contracted to conform with the time available. Suggestions for the use of any additional time available are given in paragraph 18 of the subject schedule.

Section I. GENERAL

1. PURPOSE. This subject schedule is a guide to the unit commander, operations officer, or individual instructor in preparing training schedules or units of instruction. Its use is recommended but not required.

2. TRAINING OBJECTIVE. The objective of the training outlined in this subject schedule is to acquaint the soldier with the basic functions and responsibilities of each crew member of the M26 and M45 tanks.

3. SUBJECT SCOPE. This schedule covers drills for crew drill, service of the piece, mounted action, dismounted action, evacuation of wounded from tanks, and inspections and maintenance; also, instruction in crew composition and formations, crew control, sight adjustment, destruction of equipment, and stowage.

4. TEXT REFERENCES. FM 5-25, FM 17-74; TM 9-735.

5. TRAINING AIDS. Unit commanders will find it helpful to prepare individual charts for tank crewmen, outlining their individual duties for each specific position within the tank.

6. FACILITIES AND EQUIPMENT NEEDED. a. Facilities.
    (1) A classroom.
    (2) A tank park.
    (3) An outdoor training area.

b. Equipment.
    (1) One M26 or M45 tank (stowed and equipped, less personal equipment, ammunition, and rations) per 5 students, plus one for demonstration purposes.
    (2) A loudspeaker system (if more than 5 crews are to be trained).
    (3) One tank trainer.

7. GENERAL TRAINING NOTES. a. Size of class, for maximum efficiency, should not exceed normal tank platoon size (25 men). An assistant instructor for each tank crew can be employed to advantage.

    b. Much of the initial training in crew drill can be done most efficiently through the use of trained demonstration crews.

    c. Periods 3 and 5 should be scheduled so that not more than an hour of practical work is given in one period; other instruction should be scheduled alternately with practical work in order to sustain the interest of the students.

Section II. SUBJECT SCHEDULE

8. TANK CREW DRILL, SERVICE OF THE PIECE, AND STOWAGE.

Total hours: 31

Period	Hours	Lessons	Text references	Area	Training aids and equipment
1	1½	Crew composition and formations; crew control.	Secs. II, III.	Classroom, tank park.	For each 5 men: 1 M26 (M45) tank. For instructor: 1 SCR-508 complete with mounting base, microphones, headsets, chest sets, extension cords, and antenna; 1 blackboard; 1 AN/VRC-3 radio set.
2	1	Crew drill.	Sec. IV.	Tank park.	For each 5 men: 1 M26 (M45) tank; 5 charts of duties. For instructor: Loudspeaker system.

Period	Hours	Lessons	Text references	Area	Training aids and equipment
3	3	Service of the piece and mounted action.	Secs. V, VI.	Classroom, tank park.	For each 5 men: 1 M26 (M45) tank; 5 charts of duties. For instructor: 1 tank trainer; 2 dummy rounds for cannon; 1 bell rammer and staff.
4	½	Sight adjustment.	Sec. X.	Classroom.	For instructor: 1 tank trainer; blackboard.
5	3	Dismounted action.	Sec. VII.	Tank park or training area.	For each 5 men: 1 M26 (M45) tank; 5 charts of duties. For instructor: Loudspeaker system.

96

6	1	Evacuation of wounded from tanks.	Sec. VIII.	Tank park or training area.	For each 5 men: 1 M26 (M45) tank. For instructor: Loudspeaker system.
7	16	Inspections and maintenance.	Sec. IX; and TM 9–735.	Tank park.	For each 5 men: 1 M26 (M45) tank; 5 charts of duties.
8	2	Destruction of equipment.	Sec. XI; and FM 5–25.	Classroom, tank park.	For instructor: 1 M26 (M45) tank.
9	3	Stowage.	Sec. XII; and TM 9–735.	Tank park.	For each 5 men: 1 M26 (M45) tank. For instructor: 1 M26 (M45) tank, stowed for demonstration of proper stowage.

Note. All tanks listed in *Training Aids and Equipment* column must be stowed and equipped. For the stowage demonstration in the 9th period, personal equipment, rations, and ammunition should be included.

9. FIRST PERIOD (1½ hrs.). **a. Lesson objective.** The purpose of this lesson is to familiarize tank personnel with the composition and posts of the tank crew, the operation of the interphone and radio, and the basic interphone language employed within the tank.

b. Instruction. Instruction for this period should be in the form of a 1-hour conference followed by one-half hour of practical work by the students inside the tanks, under the supervision of trained assistant instructors.

10. SECOND PERIOD (1 hr.). **a. Lesson objective.** The purpose of this lesson is to familiarize tank personnel with the basic motions of mounting and dismounting from a tank. A drill routine has been established so that these actions may become automatic.

b. Instruction. Instruction for this period should be in the form of a demonstration by a trained crew, followed by practical work on the tanks by the students. This type of instruction can be controlled best by use of the loudspeaker.

11. THIRD PERIOD (3 hrs.). **a. Lesson objective.** The purpose of this lesson is to familiarize tank personnel with the positions of the gun (howitzer) crew, the duties in operation of the tank gun (howitzer), action in case of malfunctions, and the duties of the crew members in mounted action.

b. Instruction. The first hour of this period should consist of a lecture by the instructor, followed by a demonstration of gun crew drill and mounted action by an experienced crew in a tank trainer. The remaining 2 hours should consist of practical work on the tanks. If possible, a trained assistant instructor should supervise the instruction of each tank.

12. FOURTH PERIOD (½ hr.). a. Lesson objective. The purpose of this lesson is to familiarize tank personnel with a method of rapidly checking tank sights in the field.

b. Instruction. Instruction for this period should be in the form of a conference. At this time, a review should be conducted, covering sight alinement by boresighting as outlined in FM 23-100.

13. FIFTH PERIOD (3 hrs.). a. Lesson objective. The purpose of this lesson is to familiarize tank personnel with the functions of the members of a tank crew in dismounted action.

b. Instruction. Instruction for this period should consist of a ½-hour conference on the use of dismounted action, followed by 2½ hours of instruction, drill, and practical work on the tanks. The drill portion of this training can be controlled best by using a loudspeaker.

14. SIXTH PERIOD (1 hr.). a. Lesson objective. The purpose of this lesson is to acquaint personnel with the individual duties of tank crewmen in evacuating wounded from tanks. On completion of this instruction, each man should know a fast, workable method of removing crewmen from any position in the tank.

b. Instruction. Instruction for this period should be in the form of a 15-minute lecture and a 15-minute demonstration by a trained crew, followed by one-half hour of practical work on the tanks. The position of the wounded man should be rotated throughout the tank so that crewmen may learn a method or drill applicable to each hatch of the tank.

15. SEVENTH PERIOD (16 hrs.). a. Lesson objective. The purpose of this lesson is to familiarize tank crew members with their individual duties in the performance of inspections and maintenance of personal equipment and weapons, vehicle equipment and weapons, and mechanical functions of the tank.

b. Instruction. Instruction for this period should be divided into phases and allotted throughout the training period so that vehicles can be maintained while in use. Each phase should be preceded by a 15-minute conference by the unit motor officer. The individual drills must be supervised closely by unit maintenance personnel with one mechanic per vehicle, if practicable, for the initial instruction.

16. EIGHTH PERIOD (2 hrs.). a. Lesson objective. The purpose of this lesson is to instruct tank personnel in the destruction of the tank, its weapons, and its equipment. No deliberate process has been established for the destruction of this equipment; therefore, only the theory of destruction can be taught.

b. Instruction. Instruction for this period should deal mainly with acquainting the crewmen with methods of destruction. If equipment is available, instruction in the actual use of TNT and incendiary grenades should be included. The conference and demonstration method of instruction will be found most applicable in this period.

17. NINTH PERIOD (3 hrs.). a. Lesson objective. The purpose of this lesson is to familiarize tank personnel with the proper stowage of the tank. Each man should know the location of each piece of equipment on and in the tank.

b. Instruction. Instruction for this period should begin with a ½-hour demonstration of a tank properly stowed and equipped for combat service. The demonstration should be followed by 2½ hours of practical work on the tanks. This work should consist of practice in proper stowage and restowage of a vehicle, to include the stowage of rations, fuel, ammunition and equipment.

18. REVIEW PERIODS. a. General. It should be noted that the schedule set forth herein, if rigidly followed, will attain *minimum* proficiency in the performance of tank crew duties. In order for more thorough proficiency to be attained, more extensive training is required.

b. Drills. Increased proficiency is most desirable in the exercise of crew control and the performance of drills for service of the piece, evacuation of wounded, and mounted action. The drills for inspections and maintenance should be subordinated later to actual performance of maintenance.

APPENDIX II

STOWAGE LIST

MEDIUM TANK, M26

Nomenclature	Per major item	Where carried
Section I. ORGANIZATIONAL AND VEHICULAR SPARE PARTS		
GROUP 05—COOLING SYSTEM.		
BELT, fan	2	In right front fender box.
GROUP 06—ELECTRICAL SYSTEM.		
LAMP, electric, incandescent, 24- to 28 volt, No. 1251, 3 candle-power.	8	In metal lamp box instrument panel.
LAMP, electric, incandescent, 24- to 28 volt, No. 623, 6 candle-power.	4	Do.
LAMP, electric, incandescent, 24- to 28-volt, No. 1244, 15 candle-power.	1	Do.

GROUP 13—TRACKS AND SUSPENSION.

For Vehicles equipped with T80E1 Track

BOLT, track shoe center guide	4	In right front fender box.
CONNECTION, track shoe	8	Do.
KIT, track shoe center guide	4	Do.
NUT, safety, steel, ⅞–14NF–3	4	Do.
NUT, safety, steel, ⅝–18NF–3	8	Do.
SHOE,* track, assembly (solid link)	4	On outside left turret wall in bracket.
WEDGE, track shoe connection	8	In right front fender box.

For Track T84E1 when vehicle so equipped

BOLT, track shoe center guide	4	In right front fender box.
CONNECTION, track shoe	8	Do.
KIT, track shoe center guide	4	Do.
NUT, safety, steel, ⅝–18NF–3	8	Do.
NUT, safety, steel, ⅞–14NF–3	4	Do.
SHOE,* track, assembly (rubber)	4	On outside turret left wall in bracket.
WEDGE, track shoe connection	8	In right front fender box.

GROUP 22—ACCESSORIES.

LAMP, electric, incandescent, 3-volt, No. 323	10	In metal lamp box, instrument panel tray.

* Either rubber or metal track is mounted (brackets fit only metal track; must be modified for rubber track).

MEDIUM TANK, M26—Continued

Nomenclature	Per major item	Organizational allowances — Where carried
Section II. TOOLS AND EQUIPMENT		
COMMON TOOLS.		
AXE, handled, chopping, single bit, standard grade, weight 4 pounds.	1	In right center fender box.
BAG, tool, empty.	1	In left center fender box.
BAR, crow, pinch point, diameter of stock, 1¼ inches, length 5 feet, weight 18 pounds.	1	In right center fender box.
BAR, socket wrench, cross, round, solid, diameter 7/16 inch, length 8 inches.	1	In tool bag, left center fender box.
BAR, socket, wrench, extension, ½ inch, square drive, nominal length 5 inches.	1	Do.
BAR, socket wrench, extension, ½ inch, square drive, nominal length 10 inches.	1	Do.
BAR, socket wrench, extension, ¾ inch, square drive, length 4⅝ inches, with ⅞ inch hold for pin handle.	1	Do.

BAR, socket wrench, extension, ¾ inch, square drive, nominal length 16 inches.	1	Do.
CABLE, towing, steel, diameter 1⅛ inches with 2 eyes, 1¼ by 3¼ inches, length 20 feet.	1	Coiled on rear of hull.
CHISEL, machinist's hand, cold, steel, width of cut ¾ inch, length 8 inches.	1	In tool bag, left center fender box.
CORD, light extension, 24 volt, length 15 feet	1	Do.
FILE, hand, cut smooth, length point to shoulder 10 inches	1	Do.
FILE, three-square, cut smooth, length point to shoulder 6 inches	1	Do.
HAMMER, machinist's, ball peen, weight 2 pounds	1	Do.
HANDLE, mattock, length 36 inches	1	In left center fender box.
HANDLE, socket wrench, hinged, ½ inch, square drive, length 18 inches.	1	In tool bag, left center fender box.
HANDLE, socket wrench, ½ inch, square drive, length 9½ inches minimum, 15 inches maximum.	1	Do.
HANDLE, socket wrench, speeder, brace type, ½ inch, square drive, length 12 inches.	1	Do.
HANDLE, socket wrench, "T" sliding, ½ inch, square drive, minimum length 9 inches.	1	Do.
HANDLE, socket wrench "T" sliding, ¾ inch, square drive, length 17 inches.	1	Do.
JOINT, socket wrench, universal, ½ inch, square drive	1	Do.

MEDIUM TANK, M26—Continued

Section II. TOOLS AND EQUIPMENT—Continued

Nomenclature	Per major item	Organizational allowances — Where carried
COMMON TOOLS—Continued		
MATTOCK, pick, without handle	1	In left center fender box.
PLIERS, combination, slip joint, with cutter, nominal size 6 inches	1	In tool bag, left center fender box.
PLIERS, lineman's, side cutting, length 8 inches	1	Do.
SCREWDRIVER, common, special purpose, length of blade 1½ inches, length over-all 4 inches.	1	Do.
SCREWDRIVER, machinist's, wood insert handle, length of blade 5 inches, width of blade ½ inch.	1	Do.
SHOVEL, general purpose, D-handle, round point	1	In left center fender box.
SLEDGE, blacksmith's, double face, weight 10 pounds	1	In right center fender box.
WRENCH, adjustable, single, open end, jaw opening 15/16 inch, length 8 inches.	1	In tool bag, left, center box.

WRENCH, adjustable, single, open end, jaw opening 15/16 inches, length 12 inches.	1	Do.
WRENCH, center guide bolt, socket detachable, 3/4 inch square drive, 12 point, size of opening 1 1/4 inches.	1	Do.
WRENCH, engineer's, angle 15 degrees, double end, size of openings from 5/16 and 3/8 inch to 15/16 and 1 inch.	5	Do.
WRENCH, engineer's, angle 15 degrees, double end, size of openings 9/16 and 11/16 inch.	1	Do.
WRENCH, plug, straight, size of plug 9/16 inch.	1	Do.
WRENCH, plug, straight, size of plug 3/4 inch.	1	Do.
WRENCH, set or cap screw, plug type, diameter from 3/32 to 5/8 inch, set screw size from 1/8 to 3/4 inch, cap screw size from No. 5 to 1 inch. (7 in set)	7	Do.
WRENCH, socket, detachable, 1/2 inch, square drive, 8 point, size of opening 3/8 inch.	1	In tool bag, left center fender box.
WRENCH, socket, detachable, 1/2 inch, square drive, 12 point, size of opening from 7/16 to 1 1/8 inches. (2 of 2 sizes—i. e. (2) 7/8 inches and (2) 15/16 inches.)	12	Do.
WRENCH, socket, detachable, 3/4 inch, square drive, 12 point, size of opening 1 1/2 inches.	1	Do.
SPECIAL TOOLS.		
FIXTURE, track connecting and link pulling, left-hand, right-hand (in pairs). (For use with T80E1, and T84E1 rubber backed track.)	1	In outside brackets on turret sides.

MEDIUM TANK, M26—Continued

Section II. TOOLS AND EQUIPMENT—Continued

Nomenclature	Per major item	Where carried
SPECIAL TOOLS—Continued		
WRENCH, engineer's, angle 15 degrees, single open end, size of opening 5 inches, length 42 inches.	1	Left center fender box.
WRENCH, track slack adjusting, hook spanner, diameter of circle 7⅜ inches, length 13½ inches.	1	Do.
EQUIPMENT.		
ADAPTER, gun, lubrication, hydraulic to push type, thin stem, with locking sleeve.	1	In tool bag, left center fender box.
BOX, battery and bulb stowage, assembly	1	In instrument panel tray.
BOX, flares	1	Under turret platform atop battery box.
BOX, spare parts, gun	1	Do.
BUCKET, watering, canvas, capacity 18 quarts	1	In left rear fender box.

CAN, water, capacity 5 gallons	2	One in each side center fender box.
COVER, azimuth indicator	1	On indicator.
COVER, headlamp	2	On headlamps.
EXTENSION, lubrication gun, hose type, hydraulic to hydraulic, length 12 inches.	1	In tool bag, left center fender box.
FORM, DAAGO No. 478, MWO and major unit assembly replacement record with ENVELOPE DAAGO Form No. 478–1.	1	In manual box on right of driver's seat.
GUN, lubrication, chassis, hand lever operated, capacity 15 ounces	1	In tool bag left center fender box.
HOOK, towing cable	1	In left front fender box.
OILER, pump, capacity 1 pint, length of spout 9 inches	1	In back of battery box in rack.
PADLOCK, 1½-inch	7	One each on six fender boxes, one on loader's hatch.
PAMPHLET, ORD 7, SNL G–226, Organizational Spare Parts and Equipment.	1	In manual box on right of driver's seat.
PAULIN, canvas, 12- x 12-feet	1	Strapped on left fender.
RACK, stowage rations	1	Under turret platform atop battery box.
TAPE, adhesive, 4 inches wide, 15 yards long	1	In right front fender box

MEDIUM TANK, M26—Continued

Nomenclature	Per major item	Where carried
		Organizational allowances
EQUIPMENT—Continued		
TAPE, friction, general use, black, width ¾-inch (60 feet roll)	1	In tool bag left center fender box.
## Section II. TOOLS AND EQUIPMENT—Continued		
EQUIPMENT ISSUED BY OTHER TECHNICAL SERVICES.		
APPARATUS, decontaminating, 1½ quart, M2	1	In right front fender box.
BATTERY, dry, flashlight, 1 cell	28	8 in flashlights, 8 in instruments, 12 in metal box, instrument panel tray.
CAN, 4½ ounce, flash-burn prevention cream	1	In box provided to the rear of loader in turret bulge.
CANTEEN, M1941 with CUP and COVER, M1910	5	Three in turret bulge under SCR-300 radio,

CUTTER, wire, M1938, with CARRIER	1	two in driver's compartment on brackets.
	1	In right front fender box.
EXTINGUISHER, fire, carbon dioxide, hand, capacity 4 pounds, trigger valve operated.	2	One in driver's compartment in bracket, one in turret bulge in bracket.
FLAG SET, M238	1	In turret bulge on top of radio in brackets provided.
FLASHLIGHT, electric, hand, 2 cell, with lamp	4	Two on turret wall brackets, two on steel post between driver's compartment in brackets.
KIT, first aid	1	In case inside in driver's compartment.
LAMP (spares for flashlights)	3	In metal box, instrument panel tray.
MITTENS, asbestos, M1942	2	In turret oddment box.
NET, camouflage, 45- x 45-foot	1	Strapped on right fender.
PANEL SET	1	Right center fender box.
RATIONS, field, individual type (for 5 men for 3 days).	5	50 percent in compartment under turret platform, 50 percent in right rear fender box.

MEDIUM TANK, M26—Continued

Nomenclature	Per major item	Organizational allowances Where carried
Section II. TOOLS AND EQUIPMENT—Continued		
EQUIPMENT ISSUED BY OTHER TECHNICAL SERVICES—Continued		
ROLL, blanket	5	2 in rear fender box right side 3 in rear fender box left side.
SIGNAL, ground, (assorted) (cluster or parachute; amber star, green star, or white star).	12	In flare box under turret platform atop battery box.
STOVE, cooking, 1 burner	2	In right center fender box.
TUBE, flexible nozzle, cam type	2	One in left center fender box, one in right center fender box.

Section III. ARMAMENT

GUN, 90-mm, M3	1	Mounted with gun shield on front of turret.
GUN, machine, caliber .30, Browning, M1919A4, flexible	2	One mounted in assistant driver's compartment, one coaxial mounted with 90-mm gun in mounts provided.
GUN, machine, caliber .50, Browning, M2, heavy barrel, flexible	1	On top center, rear of turret on mount provided.
MOUNT, ball, caliber .30	1	Mounted in assistant driver's compartment.
MOUNT, combination gun, M67	1	Mounted with 90-mm coaxial.
MOUNT, machine gun, AA, caliber .50	1	On top center rear of turret.
MOUNT, tripod, machine gun, caliber .30, M2	1	In left front fender box.
GUN, sub-machine, caliber .45, M3	1	In turret bulge bracket.
CARBINE, caliber .30, M2	1	Mount in bracket on left side of radio mount.
LAUNCHER, grenade, M8	1	Mount on carbine.

MEDIUM TANK, M26—Continued

Nomenclature	Per major item	Where carried
Section IV. ARMAMENT SPARE PARTS		
GUN, 90-MM, M3.		
MECHANISM, percussion, assembly	1	In spare parts box, atop battery box.
MOUNT, COMBINATION GUN, M67.		
GASKET, copper, ⅛ inch thick	2	Do.
GASKET, copper, 0.088 inch thick	2	Do.
PLUG, recoil cylinder (front end)	2	Do.
PLUG, recoil cylinder (rear end)	2	Do.
GUN, MACHINE, CALIBER .50, BROWNING, M2, NB (FLEXIBLE).		
BAR, trigger	1	In gun spare parts box.
BARREL, assembly	1	In right center fender box.
BOLT, alternate feed, assembly	1	In gun spare parts box.
COVER, assembly	1	Do.

GUN, MACHINE, CALIBER .30, BROWNING, M1919A4 (FLEXIBLE).

BARREL	2	In cover rear of ass't driver.
BOLT, assembly	1	In gun spare parts box.
COVER, assembly	1	Do.
TRIGGER	1	Do.

Section V. ARMAMENT TOOLS AND EQUIPMENT

GUN, 90-MM, M3.

BRUSH, bore, 90-mm, M19	1	In right center fender box.
BRUSH, cleaning, bristle	1	In right center fender box.
COVER, canvas, brush, bore, M518	1	In original package. Mounted on brush.
COVER, gun book, M539	1	In manual box on right side of driver's seat.
EYEBOLT, breechblock removing, length 16 inches	2	In oddment tray, turret bulge.
FORM, Department of Army, artillery gun book, O. O. Form 5825 (blank)	1	In manual box on right side of driver's seat.
HEAD, rammer, length 10⅛ inches	1	In right center fender box.
RING, wiper	1	Do.

115

MEDIUM TANK, M26—Continued

Nomenclature	Per major item	Where carried
Section V. ARMAMENT TOOLS AND EQUIPMENT—Con.		
GUN, 90-MM, M3—Continued		
SIGHT, bore, breech	1	In 90-mm gun parts box under turret platform, atop battery box.
SIGHT, bore, muzzle	1	Do.
STAFF-SECTION, end (55⅝ inches long)	1	In right center fender box.
STAFF-SECTION, middle (48¼ inches long)	3	Do.
TOOL, breechblock removing	1	In oddment tray, turret bulge.
TOOL, extracting and ramming	1	Left side of turret.
WRENCH, fuse, M7A1	1	In oddment box, turret bulge.
or		
WRENCH, fuse, M18.		

WRENCH, fuse, M16	1	Do.
WRENCH, tubular, single end, pronged	1	Do.
MOUNT, COMBINATION GUN, M67.		
BAG, empty cartridge, caliber .30	1	On mount.
BAG, and adapter ass'y. clip	1	In coaxial gun cal. .30.
COVER, combination breech and empty shell case	1	On breech.
COVER, muzzle	1	On muzzle.
COVER, receiver, caliber .30 (turret)	1	On coaxial caliber .30 gun.
GUN, oil, recoil, hand-operated	1	In 90-mm gun spare parts box under turret platform, atop battery box.
GUN, MACHINE, CALIBER .50, BROWNING, M2, HB (FLEXIBLE).		
BRUSH, cleaning caliber .50, M4	3	Gun spare parts box.
CASE, cleaning rod, M15	1	In gun spare parts box.
COVER, spare barrel, cal. .50	1	On spare barrel.
EXTRACTOR, ruptured cartridge, caliber .50	1	Gun spare parts box.
GAGE, headspace and timing	1	Do.
ROD, cleaning, jointed, caliber .50, M7	1	In gun spare parts box.
ROLL, spare parts, M14	1	Do.
ROLL, tool, canvas, empty, M10	1	Do.
WRENCH, muzzle gland and adjusting screw	1	Do.

MEDIUM TANK, M26—Continued

Section V. ARMAMENT TOOLS AND EQUIPMENT—Con.

Nomenclature	Per major item	Organizational allowances — Where carried
MOUNT, MACHINE GUN, AA, CALIBER .50		
COVER, machine gun, caliber .50	1	On AA cal. .50, gun.
GUN, MACHINE, CALIBER, .30, BROWNING, M1919A4 (FLEXIBLE).		
BRUSH, cleaning, caliber .30, M3 (wire)	6	In gun spare parts box.
BRUSH, cleaning, chamber, M6 (bristle)	1	Do.
CASE, cleaning rod, caliber .30, M1	2	Do.
CASE, cover group	1	Do.
CASE, spare bolt, M2	1	Do.
COVER, spare barrel	2	On spare barrels.
EXTRACTOR, ruptured cartridge	2	One for coaxial in odd-ment tray; one for assistant driver, hooked to gun mount.

118

ROD, cleaning, jointed, caliber .30, M1	2	In gun spare parts box.
ROLL, spare parts, M13	1	Do.
SCREWDRIVER, common, normal duty, single grip, length over-all 7 inches.	1	Do.
WRENCH, barrel bearing plug	1	Do.
WRENCH, combination, M6	1	Do.
MOUNT, BALL CALIBER .30.		
BAG, empty, cartridge, caliber .30	1	On caliber .30 bow gun.
BAG, and adapter ass'y clip	1	Do.
COVER, muzzle, caliber .30	1	Do.
COVER, receiver, caliber .30	1	Do.
MOUNT, TRIPOD, MACHINE GUN, CALIBER .30, M2.		
HOOD, tripod mount, caliber .30	1	On tripod in left front fender box.
GUN, SUB-MACHINE, CAL. .45, M3A1.		
CASE, gun, sub-machine, cal. .45, M3 clip	1	In turret bulge.

Section VI. SIGHTING AND FIRE CONTROL EQUIPMENT

COVER, telescope	1	On telescope.
HEAD, assy. (spares) (furnished only w/periscope M6)	2	In spare periscope box.
HEADREST, assembly	1	Mounted on gunner's periscope.

MEDIUM TANK, M26—Continued

Section VI.—SIGHTING AND FIRE CONTROL EQUIPMENT—Continued

Nomenclature	Per major item	Where carried
HOLDER, periscope (only supplied when periscope M13 or M6 is furnished Commander).	1	Mounted on the Commander's hatch.
LAMP, electric, incandescent, 3 volt, No. 323 (for azimuth indicator)	5	In metal box, on instrument panel tray.
MOUNT, periscope, M73 (T113)	1	On gunner's periscope.
MOUNT, telescope, M72 (T90)	1	On 90-mm gun mount.
PERISCOPE, M13 or PERISCOPE, M6.	5	One in holder, driver position; one in holder, bow gun position; one in holder, loader position; one spare in instrument panel tray; one spare under 90-mm gun.

120

PERISCOPE, M15 or PERISCOPE, M13, or PERISCOPE, M6.	1	One in holder, tank commander's position.
PERISCOPE, M16 or PERISCOPE, M10F.	2	One in holder, gunner's position; one spare in turret bulge, under radio mount.
QUADRANT, elevation, M9	1	On 90-mm gun.
QUADRANT, gunner's, M1	1	On right side turret wall in bracket provided.
SETTER, fuse, M27 (wrench) or SETTER, fuse, M14 (wrench)	1	Gun spare parts box.
SETTER, fuse, M26 or SETTER, fuse, M22 or SETTER, fuse, M23	1	In turret bulge.
	1	Do.
	1	Do.

MEDIUM TANK, M26—Continued

Nomenclature	Per major item	Where carried
Section VI. SIGHTING AND FIRE CONTROL EQUIPMENT—Continued		
TELESCOPE, M83C	1	In telescope mount on right side of 90-mm gun.
or		
TELESCOPE, M71N,		
or		
TELESCOPE, M71C.		
Section VII. AMMUNITION		
GRENADE, hand, fragmentation	4	In box on left of tank commander's seat on turret ring.
GRENADE, hand, smoke, M15	4	Do.
GRENADE, hand, smoke, M16 or M18	4	Do.

ROUNDS, for 90-mm gun	70	60 in floor racks; 10 in ready rack on left side of turret wall.
ROUNDS, for caliber .30 machine gun (in metallic link belts)	5,500	Turret hull and gun as marked.
ROUNDS, for caliber .50 machine gun	550	Turret bulge and gun as marked.
ROUNDS, for caliber .45 submachine gun	180	In turret bulge rack and on right of driver's seat in cases.
ROUNDS, for caliber .30 carbine	90	One clip in carbine, two clips in bracket near carbine.

Section VIII. T/O & E EQUIPMENT

(T/O & E's 17-27N, 17-36N, and 17-73N)

BAG, carrying, ammunition	1	Stowed in conjunction with ammunition on right of driver's seat.
BINOCULAR, M17A1	1	On turret wall, on tank commander's side.
CASE, carbine, caliber, .30, M2	1	On carbine.

MEDIUM TANK, M26—Continued

Nomenclature	Per major item	Where carried
Section VIII. T/O & E EQUIPMENT—Continued		
MARKER, luminous, radioactive, type 1	2	In compartment in right front fender box.
MASK, gas, M9	1	Attached to crew member's tank seat, using hooks provided for stowage.
RADIO SET, AN/VRC-3	1	Mounted in turret bulge.
RADIO SET, SCR-508	1	Mounted in radio bracket.
Chest, CH-264, spare parts	1	Stowed behind SCR-508 transmitter.
Roll, BG-56-A, with spare mast sections, MS 116, 117, and 118	1	Do.
RADIO SET, SCR-528	1	Mounted in radio bracket (per tank other than those authorized SCR-508).

Chest, CH-264, spare parts	1	Stowed behind SCR-528 transmitter.
Roll, BG-56-A, with spare mast sections, MS-116, 117, and 118	1	Stowed behind SCR-528 transmitter.
RESPIRATOR, dust, M2	1	Looped around carrying case of gas mask.
TABLE, graphic firing, M42	1	In manual box.

Section IX. INDIVIDUAL OR ORGANIZATIONAL EQUIPMENT ALLOWANCES

(T/A 21.)

BAG, sleeping, wool, M1945	5	When authorized, same as blanket roll.
CASE, water repellent, bag, sleeping, M1945	5	Do.
LINER, helmet, crash, M1	5	On person.
PACK, field cargo and combat, M1945 (with individual clothing as prescribed by unit Commander)	5	Center fender boxes.
RAINCOAT, rubberized, M1938 (dismounted)	5	Attached to field pack.

NOTES

1. All references to right and left are based on individual standing in rear of tank, facing direction of forward movement, with gun pointed to the front.
2. See figures in TM 9-735 for further information as to the location of the items of equipment.

APPENDIX III

STOWAGE LIST

MEDIUM TANK, M45

Section I. ORGANIZATIONAL AND VEHICULAR SPARE PARTS

Nomenclature	Organizational allowances	
	Per major item	Where carried
GROUP 05—COOLING SYSTEM.		
BELT, fan _____	2	In right front fender box.
GROUP 06—ELECTRICAL SYSTEM.		
LAMP, electric, incandescent, 24- to 28-volt, No. 1251, 3 candlepower.	8	In metal lamp box instrument panel.
LAMP, electric, incandescent, 24- to 28-volt, No. 623, 6 candlepower.	4	Do.

126

LAMP, electric, incandescent, 24- to 28-volt, No. 1244, 15 candle-power.	1	In metal lamp box instrument panel.
GROUP 13—TRACKS AND SUSPENSION. *For Vehicles equipped with T80E1 Track.*		
BOLT, track shoe center guide	4	In right front fender box.
CONNECTION, track shoe	8	Do.
KIT, track shoe center guide	4	Do.
NUT, safety, steel, 7/8–14NF–3	4	Do.
NUT, safety, steel, 5/8–18NF–3	8	Do.
Shoe,* track, assembly (solid link)	4	On outside left turret wall in bracket.
WEDGE, track shoe connection	8	In right front fender box.
For Track T84E1 when vehicle so equipped		
BOLT, track shoe center guide	4	Do.
CONNECTION, track shoe	8	Do.
KIT, track shoe center guide	4	Do.
NUT, safety, steel, 5/8–18NF–3	8	Do.
NUT, safety, steel, 7/8–14NF–3	4	Do.
SHOE,* track, assembly (rubber)	4	On outside turret left wall in bracket.
WEDGE, track shoe connection	8	In right front fender box

* Either rubber or metal track is mounted (brackets fit only metal track; must be modified for rubber track).

MEDIUM TANK, M45—Continued

Nomenclature	Organizational allowances	
	Per major item	Where carried

Section I. ORGANIZATIONAL AND VEHICULAR SPARE PARTS—Continued

GROUP 22—ACCESSORIES.

LAMP, electric, incandescent, 3-volt No. 323	10	In metal lamp box, instrument panel tray.

Section II. TOOLS AND EQUIPMENT

COMMON TOOLS.

AXE, handled, chopping, single bit, standard grade, weight 4 pounds.	1	In right center fender box.
BAG, tool, empty	1	In left center fender box.
BAR, crow, pinch point, diameter of stock, 1¼ inches, length 5 feet, weight 18 pounds.	1	In right center fender box.
BAR, socket wrench, cross, round, solid, diameter 7/16 inch, length 8 inches.	1	In tool bag, left center fender box.

BAR, socket, wrench, extension, ½ inch, square drive, nominal length 5 inches.	1	Do.
BAR, socket wrench, extension, ½ inch, square drive, nominal length 10 inches.	1	Do.
BAR, socket wrench, extension, ¾ inch, square drive, length 4⅝ inches, with ⅞ inch hold for pin handle.	1	Do.
BAR, socket wrench, extension, ¾ inch, square drive, nominal length 16 inches.	1	Do.
CABLE, towing, steel, diameter 1⅛ inches with 2 eyes, 1¼ by 3¼ inches, length 20 feet.	1	Coiled on rear of hull.
CHISEL, machinist's hand, cold, steel, width of cut ¾ inch, length 8 inches.	1	In tool bag, left center fender box.
CORD, light extension, 24 volt, length 15 feet	1	Do.
FILE, hand, cut smooth, length point to shoulder 10 inches	1	Do.
FILE, three-square, cut smooth, length point to shoulder 6 inches	1	Do.
HAMMER, machinist's, ball peen, weight 2 pounds	1	Do.
HANDLE, mattock, length 36 inches	1	In left center fender box.
HANDLE, socket wrench, hinged, ½ inch, square drive, length 18 inches.	1	In tool bag, left center fender box.
HANDLE, socket wrench, ½ inch, square drive, length 9½ inches minimum, 15 inches maximum.	1	Do.
HANDLE, socket wrench, speeder, brace type, ½ inch, square drive, length 12 inches.	1	Do.

MEDIUM TANK, M45—Continued

Nomenclature	Per major item	Organizational allowances — Where carried
Section II. TOOLS AND EQUIPMENT—Continued		
COMMON TOOLS—Continued		
HANDLE, socket wrench, "T" sliding, ½ inch, square drive, minimum length 9 inches.	1	In tool bag, left center fender box.
HANDLE, socket wrench "T" sliding, ¾ inch, square drive, length 17 inches.	1	Do.
JOINT, socket wrench, universal, ½ inch, square drive	1	Do.
MATTOCK, pick, without handle	1	In left center fender box.
PLIERS, combination, slip joint, with cutter, nominal size 6 inches.	1	In tool bag, left center fender box.
PLIERS, lineman's, side cutting, length 8 inches	1	Do.
SCREWDRIVER, common, special purpose, length of blade 1½ inches, length over-all 4 inches.	1	Do.
SCREWDRIVER, machinist's, wood insert handle, length of blade 5 inches, width of blade ½ inch.	1	Do.

SHOVEL, general purpose, D-handle, round point	1	In left center fender box.
SLEDGE, blacksmith's, double face, weight 10 pounds	1	In right center fender box.
WRENCH, adjustable, single, open end, jaw opening $15/16$ inch, length 8 inches.	1	In tool bag, left center box.
WRENCH, adjustable, single, open end, jaw opening $1\frac{5}{16}$ inches, length 12 inches.	1	Do.
WRENCH, center guide bolt, socket detachable, $\frac{3}{4}$ inch square drive, 12 point, size of opening $1\frac{1}{4}$ inches.	1	Do.
WRENCH, engineer's, angle 15 degrees, double end, size of openings from $\frac{5}{16}$ and $\frac{3}{8}$ inch to $15/16$ and 1 inch.	5	Do.
WRENCH, engineer's, angle 15 degrees, double end, size of openings $9/16$ and $11/16$ inch.	1	Do.
WRENCH, plug, straight, size of plug $9/16$ inch	1	Do.
WRENCH, plug, straight, size of plug $\frac{3}{4}$ inch	1	Do.
WRENCH, set or cap screw, plug type, diameter from $3/32$ to $\frac{5}{8}$ inch, set screw size from $\frac{1}{8}$ to $\frac{3}{4}$ inch, cap screw size from No. 5 to 1 inch. (7 in set).	7	Do.
WRENCH, socket, detachable, $\frac{1}{2}$ inch, square drive, 8 point, size of opening $\frac{3}{8}$ inch.	1	Do.
WRENCH, socket, detachable, $\frac{1}{2}$ inch, square drive, 12 point, size of opening from $7/16$ to $1\frac{1}{8}$ inches. (2 of 2 sizes—i. e. (2) $\frac{7}{8}$ inch and (2) $15/16$ inch.)	12	Do.
WRENCH, socket, detachable, $\frac{3}{4}$ inch, square drive, 12 point, size of opening $1\frac{1}{2}$ inches.	1	Do.

MEDIUM TANK, M45—Continued

Nomenclature	Per major item	Where carried
Section II. TOOLS AND EQUIPMENT—Continued		
SPECIAL TOOLS.		
FIXTURE, track connecting and link pulling, left-hand, right-hand (in pairs) (for use with T80E1, and T84E1 rubber backed track).	1	In outside brackets on turret sides.
WRENCH, engineer's, angle 15 degrees, single open end, size of opening 5 inches, length 42 inches.	1	Left center fender box.
WRENCH, track slack adjusting, hook spanner, diameter of circle 7⅜ inches, length 13½ inches.	1	Do.
EQUIPMENT.		
ADAPTER, gun, lubrication, hydraulic to push type, thin stem, with locking sleeve.	1	In tool bag, left center fender box.
BOX, battery and bulb stowage, assembly	1	In instrument panel tray.
BOX, flares	1	Under turret platform atop battery box.

BOX, spare parts, gun	1	Do.
BUCKET, watering, canvas, capacity 18 quarts	1	In left rear fender box.
CAN, water, capacity 5 gallons	2	One in each side center fender box.
COVER, azimuth indicator	1	On indicator.
COVER, headlamp	2	On headlamps.
EXTENSION, lubrication gun, hose type, hydraulic to hydraulic, length 12 inches	1	In tool bag, left center fender box.
FORM, DAAGO No. 478, MWO and major unit assembly replacement record with ENVELOPE DAAGO Form No. 478-1.	1	In manual box on right of driver's seat.
GUN, lubrication, chassis, hand lever operated, capacity 15 ounces	1	In tool bag left center fender box.
HOOK, towing cable	1	In left front fender box.
OILER, pump, capacity 1 pint, length of spout 9 inches	1	In back of battery box in rack.
PADLOCK, 1½ inch	7	One each on six fender boxes, one on loader's hatch.
PAMPHLET, ORD 7, SNL G-226, Organizational Spare Parts and Equipment.	1	In manual box on right of driver's seat.
PAULIN, canvas, 12- x 12-feet	1	Strapped on left fender.
RACK, stowage rations	1	Under turret platform atop battery box.

MEDIUM TANK, M45—Continued

Nomenclature	Per major item	Where carried
EQUIPMENT—Continued		
TAPE, adhesive, 4 inches wide, 15 yards long	1	In right front fender box.
TAPE, friction, general use, black, width ¾ inch (60 feet roll)	1	In tool bag left center fender box.
EQUIPMENT ISSUED BY OTHER TECHNICAL SERVICES.		
APPARATUS, decontaminating, 1½-quart, M2	1	In right front fender box.
BATTERY, dry, flashlight, 1 cell	28	8 in flashlights, 8 in instruments, 12 in metal box, instrument panel tray.
CAN, 4½-ounce, flash-burn prevention cream	1	In box provided to the rear of loader in turret bulge.

Section II. TOOLS AND EQUIPMENT—Continued

134

Item		Location
CANTEEN, M1941 with CUP and COVER, M1910	5	Three in turret bulge under SCR-300 radio, two in driver's compartments on brackets.
CUTTER, wire, M1938, with CARRIER	1	In right front fender box.
EXTINGUISHER, fire, carbon dioxide, hand, capacity 4 pounds, trigger valve operated.	2	One in driver's compartment in bracket, one in turret bulge in bracket.
FLAG SET, M238	1	In turret bulge on top of radio in brackets provided.
FLASHLIGHT, electric, hand, 2 cell, with lamp	4	Two on turret wall brackets, two on steel post between driver's compartment in brackets.
KIT, first aid	1	In case inside in driver's compartment.
LAMP (spares for flashlights)	3	In metal box, instrument panel tray.
MITTENS, asbestos, M1942	2	In turret oddment box.
NET, camouflage, 45- x 45-foot	1	Strapped on right fender.
PANEL SET	1	Right center fender box.

MEDIUM TANK, M45—Continued

Section II. TOOLS AND EQUIPMENT—Continued

Nomenclature	Per major item	Organizational allowances — Where carried
EQUIPMENT ISSUED BY OTHER TECHNICAL SERVICES—Continued		
RATIONS, field, individual type (for 5 men for 3 days)	5	50 percent in compartment under turret platform, 50 percent in right rear fender box.
ROLL, blanket	5	2 in rear fender box right side. 3 in rear fender box left side.
SIGNAL, ground (assorted) (cluster or parachute; amber star, green star, or white star).	12	In flare box under turret platform atop battery box.
STOVE, cooking, 1 burner	2	In right center fender box

TUBE, flexible nozzle, cam type	2	One in left center fender box, one in right center fender box.

Section III. ARMAMENT

CARBINE, caliber .30, M2	1	In bracket on left side of radio mount.
GUN, machine, caliber .30, Browning, M1919A4, flexible	2	One mounted in assistant driver's compartment, one coaxial mounted with 105-mm howitzer.
GUN, sub-machine, caliber .45, M3	1	In turret bulge bracket.
GUN, machine, caliber .50, Browning, M2, heavy barrel, flexible	1	On top center, rear of turret on mount.
HOWITZER, 105-mm, M4	1	Mounted with gun shield on front of turret.
LAUNCHER, grenade, M8	1	On carbine.
MOUNT, ball, caliber .30	1	Mounted in assistant driver's compartment.
MOUNT, combination gun, M71 (T117)	1	Mounted with 105-mm coaxial.
MOUNT, machine gun AA, caliber .50	1	On top center rear of turret.

MEDIUM TANK, M45—Continued

Nomenclature	Per major item	Where carried
Section III. ARMAMENT—Continued		
MOUNT, tripod, machine gun, caliber .30, M2	1	In left front fender box.
STABILIZER	1	Attached between howitzer and its mount.
Section IV. ARMAMENT SPARE PARTS		
HOWITZER, 105-MM, M4.		
MECHANISM, percussion, assembly	1	In spare parts box under turret platform atop battery box.
MOUNT, COMBINATION GUN, M71 (T117).		
GASKET, copper, ⅛ inch thick	2	Do.
GASKET, copper, 0.088 inch thick	2	Do.

PLUG, recoil cylinder (front end)	2	Do.
PLUG, recoil cylinder (rear end)	2	Do.
GUN, MACHINE, CALIBER .50, BROWNING, M2, HB FLEXIBLE.		
BAR, trigger	1	In gun spare parts box.
BARREL, assembly	1	In right center fender box.
BOLT, alternate feed, assembly	1	In gun spare parts box.
COVER, assembly	1	Do.
GUN, MACHINE, CALIBER .30, BROWNING, M1919A4 (FLEXIBLE).		
BARREL	2	In cover rear of ass't driver.
BOLT, assembly	1	In gun spare parts box.
COVER, assembly	1	Do.
TRIGGER	1	Do.

Section V. ARMAMENT TOOLS AND EQUIPMENT

HOWITZER, 105-MM M4.		
BRUSH, bore, 105-mm M12	1	In right center fender box.
BRUSH, channel type, 105-MM	1	In spare parts box in original packing.
COVER, bore brush, M515	1	On brush.
COVER, gun book, M539	1	In manual box on right of driver's seat.

MEDIUM TANK, M45—Continued

Nomenclature	Per major item	Organizational allowances Where carried

Section V. ARMAMENT TOOLS AND EQUIPMENT—Con.

HOWITZER, 105-MM M4—continued

Nomenclature	Per major item	Where carried
FORM, Department of Army, artillery gun book, O.O No. 5825 (blank).	1	In manual box on right of driver's seat.
HAMMER, hide face, weight 2 pounds, diameter of face 1¾ inches	1	Gun spare parts box.
RAMMER, cleaning and unloading, M5	1	In right center fender box.
SIGHT, bore, breech	1	In gun spare parts box under turret platform atop battery box.
SIGHT, bore, muzzle	1	Do.
STAFF-SECTION, end (61⅜ inches long)	1	In right center fender box.
STAFF-SECTION, intermediate (51¹⁵⁄₁₆ inches long)	1	Do.
TOOL, crimping, 105-mm	1	Gun spare parts box.

WRENCH, fuze, M7A1 or	1	In gun parts box under turret platform atop battery box.
WRENCH, fuze, M18.		
WRENCH, fuze, M16	1	Do.
MOUNT, COMBINATION GUN, M71 (T117).		
BAG, and adapter assembly, clip	1	On caliber .30 coaxial mount.
BAG, empty cartridge, caliber .30	1	Do.
COVER, breech	1	On howitzer.
COVER, muzzle, M4	1	Do.
COVER, receiver, caliber .30 (turret)	1	On caliber .30 coaxial mount.
GUN, oil, recoil, hand-operated	1	In gun parts box under turret platform atop battery box.
GUN, MACHINE, CALIBER .50, BROWNING, M2, HB (FLEXIBLE).		
BRUSH, cleaning, caliber .50, M4	3	Gun spare parts box.
CASE, cleaning rod, M15	1	In gun spare parts box.
COVER, spare barrel, cal. .50	1	On spare barrel.
EXTRACTOR, ruptured cartridge, caliber .50	1	Gun spare parts box.
GAGE, headspace and timing	1	Do.
ROD, cleaning, jointed, caliber .50, M7	1	In gun spare parts box.

MEDIUM TANK, M45—Continued

Nomenclature	Per major item	Where carried
Section V. ARMAMENT TOOLS AND EQUIPMENT—Con.		
GUN, MACHINE, CALIBER .50, BROWNING, M2, HB (FLEXIBLE)—Continued		
ROLL, spare parts, M14	1	In gun spare parts box.
ROLL, tool, canvas, empty, M10	1	Do.
WRENCH, muzzle gland and adjusting screw	1	Do.
MOUNT, MACHINE GUN, AA, CALIBER .50.		
COVER, machine gun, caliber .50	1	On AA cal. .50, gun.
GUN, MACHINE, CALIBER .30, BROWNING, M1919A4 (FLEXIBLE).		
BRUSH, cleaning, caliber .30, M3 (wire)	6	In gun spare parts box.
BRUSH, cleaning, chamber, M6 (bristle)	1	Do.
CASE, cleaning rod, caliber .30, M1	2	Do.
CASE, cover group	1	Do.
CASE, spare bolt, M2	1	Do.

COVER, spare barrel	2	On spare barrels.
EXTRACTOR, ruptured cartridge	2	One for coaxial in odd-ment tray; one for assistant driver, hooked to gun mount.
ROD, cleaning, jointed, caliber .30, M1	2	In gun spare parts box.
ROLL, spare parts, M13	1	Do.
SCREWDRIVER, common, normal duty, single grip, length over-all 7 inches.	1	Do.
WRENCH, barrel bearing plug	1	Do.
WRENCH, combination, M6	1	Do.
MOUNT, BALL CALIBER .30.		
BAG, empty, cartridge, caliber .30	1	On caliber .30 bow gun.
BAG, and adapter ass'y clip	1	Do.
COVER, muzzle, caliber .30	1	Do.
COVER, receiver, caliber .30	1	Do.
MOUNT, TRIPOD, MACHINE GUN, CALIBER .30, M2.		
HOOD, tripod mount, caliber .30	1	On tripod in left front fender box.
GUN, SUB-MACHINE CAL. .45, M3A1.		
CASE, gun, sub-machine cal. .45, M3 clip	1	In turret bulge.

MEDIUM TANK, M45—Continued

Nomenclature	Per major item	Organizational allowances — Where carried

Section VI. SIGHTING AND FIRE CONTROL EQUIPMENT

Nomenclature	Per major item	Where carried
COVER, telescope	1	On telescope.
HEAD, ass'y (spares) (furnished only when periscope M6 is supplied)	2	In spare periscope box.
HOLDER, periscope (only furnished when periscope M13 or M6 is supplied for commander's use).	1	Mounted on tank commander's hatch.
HOLDER, telescope, mount, M4 (T3)	1	On howitzer mount.
LAMP, electric, incandescent, 3-volt, No. 323	5	In metal box, instrument panel tray.
MOUNT, periscope, M83 (T130)	1	On gunner's periscope.
MOUNT, telescope, M82 (T131)	1	On howitzer mount.
PERISCOPE, M13 or PERISCOPE, M6.	5	One in holder at driver's position, one in holder at bow gun position, one in holder at loader's position, one spare in

144

PERISCOPE, M15	1	instrument panel tray, one spare under howitzer. In holder at tank commander's position.
or		
PERISCOPE, M13,		
or		
PERISCOPE, M6.		
PERISCOPE, M16	2	One in holder at gunner's position, one spare in turret bulge on right of radio.
or		
PERISCOPE, M10D.		
QUADRANT, elevation, M9	1	On howitzer.
QUADRANT, gunner's M1	1	On left side turret wall in bracket provided.
SETTER, fuze, M27 (wrench)	1	In gun spare parts box.
or		
SETTER, fuze, M14 (wrench)	1	Do
SETTER, fuse, M26	1	Do.
or		
SETTER, fuse, M22	1	Do.
SIGHT UNIT, M29A1	1	On right side of howitzer.
TELESCOPE, M76G	1	On mount provided.

145

MEDIUM TANK, M45—Continued

Section VII. AMMUNITION

Nomenclature	Per major item	Where carried
GRENADE, hand, fragmentation	4	In box on left of tank commander's seat on turret ring.
GRENADE, hand, smoke, M15	4	Do.
GRENADE, hand, smoke, M16 or M18	4	Do.
ROUNDS, for caliber .30 machine gun, (in metallic link belts)	5,500	In turret, hull, and gun.
ROUNDS, for caliber .50 machine gun	550	In turret bulge and gun.
ROUNDS, for caliber .45 sub-machine gun	180	In turret bulge rack and on right of driver's seat, in cases.
ROUNDS, for 105-mm howitzer	74	In floor and ready racks.
ROUNDS, for caliber .30 carbine	90	One clip in carbine, two clips in case in bracket on turret bulge.

Section VIII. T/O & E EQUIPMENT

(T/O & E 17-57N.)

BAG, carrying ammunition	1	Stowed in conjunction with ammunition on right of driver's seat.
BINOCULAR, M17A1	1	On modified bracket on side of turret wall.
CASE, carbine, caliber .30, M2	1	Around carbine.
LINER, helmet, crash, M1	5	Worn on person.
MARKER, luminous, radioactive, type 1	2	In compartment in right front fender box.
MASK, gas, M9	1	Attached to crew member's tank seat using hooks provided for stowage.
RADIO SET, AN/VRC-3	1	Mounted in turret bulge.
RADIO SET, SCR-508	1	Mounted in radio bracket.
Chest, CH-264, spare parts for SCR-508	1	Stowed behind SCR-508 transmitter.
Roll, BG-56-A, with spare mast sections, MS 116, 117, and 118	1	Do.

MEDIUM TANK, M45—Continued

Section VIII. T/O & E EQUIPMENT—Continued

Nomenclature	Organizational allowances	
	Per major item	Where carried
RADIO SET, SCR-528	1	Mounted in radio bracket (per tank other than those authorized SCR-508).
Chest, CH-264, spare parts for SCR-528	1	Stowed behind SCR-528 transmitter.
Roll, BG-56-A, with spare mast sections, MS 116, 117, and 118	1	Do.
RESPIRATOR, dust, M2	1	Looped around carrying case of gas mask.
TABLE, graphic firing, M39	1	In manual box.

Section IX. INDIVIDUAL OR ORGANIZATIONAL EQUIPMENT ALLOWANCES

(T/A 21.)

BAG, sleeping, wool, M1945	5	When authorized, same as blanket roll.
CASE, water repellant, bag, sleeping, M1945	5	Do.
PACK, field, cargo and combat, M1945 (with individual's clothing as prescribed by unit commander).	5	Center fender box.
RAINCOAT, rubberized, M1938, dismounted	5	Attached to field pack.

NOTES

1. All references to right and left are based on individual standing in rear of tank, facing direction of forward movement, with howitzer pointed to the front.
2. See figures in TM 9–735 for further information as to the location of items of equipment.

APPENDIX IV

REFERENCES

SR 385-310-1	Regulations for Firing Ammunition for Training, Target Practice, and Combat.
FM 5-25	Explosives and Demolitions.
FM 21-6	List and Index of Department of the Army Publications.
FM 21-7	List of War Department Films, Film Strips, and Recognition Film Slides.
FM 21-8	Military Training Aids.
FM 23-55	Browning Machine Guns, Caliber .30, M1917A1, M1919A4, M1919A6.
FM 23-100	Tank and Tank Destroyer Gunnery.
TM 9-324	105-MM Howitzer M4 (Mounted in Combat Vehicles).
TM 9-374	90-MM Gun M3, Mounted in Combat Vehicles.
TM 9-735	Medium Tanks, M26 and M45.
TM 9-1900	Ammunition, General.
TM 9-1901	Artillery Ammunition.
TM 11-600	Radio Sets SCR-508-A, C, D, AM, CM, DM; SCR-528-A, C, D, AM, CM, DM; and AN/VRC-5.
TM 11-2758	Installation of Radio and Interphone Equipment in Tank, Medium, M26.
TM 20-205	Dictionary of United States Army Terms.

INDEX

	Paragraphs	Pages
Abandoning tank	43	50
Action in case of fire	45	51
Adjustment of sights	63	82
After-operation maintenance	61	77
Ammunition:		
Loading	40	42
Use and withdrawing	39	40
At-halt inspection	60	74
Authority for destroying equipment	64	83
Before-operation inspection, maintenance	58	63
Bivouac safety precautions	50	54
Breech, opening and closing	21, 22	21
Burning of equipment	69–71	88
Checking interphone and radio	6, 9, 10	4, 7, 8
Checking sights	63	82
Closing breech	22	21
Closing hatches	15	14
Commands, interphone	12	9
Composition of crew	3	1
Conduct of instruction, dismounted action	46	52
Conduct of training	app. I	93
Control box positions, settings	7, 8	5
Crew members, duties in firing	36	35
Designations of crew members	3	1
Destroying tank, equipment	44, 64–71	50, 83
Dismounted action, instruction	46	52
Dismounted drill	13	11
Dismounted posts formations	4	2
Dismounting from tank	16, 17	16, 17

	Paragraphs	Pages
Drill:		
Abandoning tank	43	50
Action in case of fire	45	51
Closing hatches	15	14
Dismounted	13	11
Dismounting	16, 17	16, 17
Evacuation of wounded	54–56	57
Fighting on foot	41	43
Firing	36	35
Inspections and maintenance	58–62	63
Mounting tank	14	13
Opening hatches	15	14
Pep	18	20
Preparing to fire	35	32
Remounting from action	42	48
Securing guns	37	38
During-operation inspection	59	74
Equipment, destruction	64–71	83
Evacuation of wounded:		
Drill	54–56	57
General	52	56
Methods	53	56
Extinguishing fires	45	51
Extraction, malfunctions	34	31
Failure to load, fire, extract	31–34	28
Field check of sights	63	82
Fighting on foot	41	43
Fire:		
Action in case of	45	51
Commands	12	9
Prevention safety precautions	48	53
Firing:		
Duties of crew	36	35
Gun	26	25
Malfunction	33	30
Safety precautions	23	21
Formations	4	2
Gun, loading, laying, firing	24–26	23
Guns, loading, general	38	40
Guns, securing	37	38

	Paragraphs	Pages
Halt, inspection	60	74
Hatches, opening and closing	15	14
Inspections	57–62	61
Instruction in dismounted action, conduct of	46	52
Instruction, schedules, preparation	app. I	93
Interphone, operation	5–10	3
Interphone terminology	11, 12	8, 9
Language, interphone	12	9
Laying the gun	25	24
Loading:		
Ammunition	40	42
Gun	24, 38	23, 38
Malfunction in	32	28
Lubrication, 250-mile	62	79
Maintenance	57–62	61
Malfunctions	31–34	28
Methods of destroying equipment	64–71	83
Methods of evacuating wounded	53–56	56
Miscellaneous safety precautions	51	55
Mounted posts formation	4	2
Mounting tank	14	13
Safety precautions	47	53
Movement of tank, commands	12	9
Opening breech	21	21
Opening hatches	15	14
Operating tank, safety precautions in	49	54
Operation of interphone and radio	5–10	3
Order, withdrawing ammunition	39	40
Park safety precautions	50	54
Pep drill	18	20
Periodic maintenance	62	79
Positions, crew	4, 20	2, 20
Positions, interphone control box	7	5
Precautions, safety. (*See* Safety precautions.)		
Preparing to fire	35	32
Principles of destroying equipment	64	83

153

	Paragraphs	Pages
Procedure:		
Evacuation of wounded	54–56	57
Interphone and radio check	6	4
Withdrawing ammunition	39	40
Program, training	app. I	93
Projectile, stuck, removal	28	26
Purpose of manual	1	1
Radio, operation	5–10	3
References		
Remounting from dismounted action	42	48
Removing unfired, stuck rounds, projectiles	27–30	25
Rounds, unfired, stuck, removing	27, 29, 30	25, 26, 27
Safety precautions:		
During firing	23	21
Fire prevention	48	53
Miscellaneous	51	55
Mounting and operating tank	47, 49	53, 54
Park and bivouac	50	54
Schedules, subject preparation	app. I	93
Scope of manual	2	1
Securing guns	37	38
Sequence, use of ammunition	39	40
Service of the piece, general	19	20
Setting radio-interphone switches	8	5
Sights, field check	63	82
Stowage	72–74	92
Stowage lists	app. II, III	102, 126
Stuck projectile, round, removal	27–30	25
Subject schedules, preparation	app. I	93
Switches, setting, radio-interphone	8	5
Terminology, interphone	11, 12	8, 9
Testing interphone and radio	6	4
Tools for destroying equipment	64	83
Training:		
Destroying equipment	64	83
Schedules, preparation	app. I	93
Stowage	74	92
Turret, commands for control	12	9

	Paragraphs	Pages
Unloading unfired, stuck rounds	27, 29, 30	25, 26, 27
Use of ammunition	39	40
Use of definite terminology	11	8
Weapons, loading	38	40
Weapons, personal, stowage	73	92
Weekly maintenance	62	79
Withdrawing ammunition	39	40
Wounded, evacuation of. (*See* Evacuation.)		

RESTRICTED

RESTRICTED

©2013 Periscope Film LLC
All Rights Reserved
ISBN#978-1-937684-48-8
www.PeriscopeFilm.com

www.ingramcontent.com/pod-product-compliance
Lightning Source LLC
Chambersburg PA
CBHW071718090426
42738CB00009B/1811

Mysterious Signs Of The Torah Revealed In GENESIS

Dr. Akiva Gamliel Belk

- Founder -

B'nai Noach Torah Institute, LLC

http://www.bnti.us

Mysterious Signs Of The Torah Revealed In GENESIS

Copyright © 2012 Dr. Akiva Gamliel Belk

All rights reserved

ISBN-10: 0615685390

ISBN-13: 978-0615685397